INTRODUCING ROBOTIC PROCESS AUTOMATION TO YOUR ORGANIZATION

A GUIDE FOR BUSINESS LEADERS

Robert Fantina
Andriy Storozhuk
Kamal Goyal

Apress®

Introducing Robotic Process Automation to Your Organization: A Guide for Business Leaders

Robert Fantina
Kitchener, ON, Canada

Andriy Storozhuk
London, ON, Canada

Kamal Goyal
Kitchener, ON, Canada

ISBN-13 (pbk): 978-1-4842-7415-6   ISBN-13 (electronic): 978-1-4842-7416-3
https://doi.org/10.1007/978-1-4842-7416-3

Managing Director, Apress Media LLC: Welmoed Spahr
Acquisitions Editor: Shivangi Ramachandran
Development Editor: Laura Berendson
Coordinating Editors: Rita Fernando and Mark Powers

Cover designed by eStudioCalamar

Distributed to the book trade worldwide by Apress Media, LLC, 1 New York Plaza, New York, NY 10004, U.S.A. Phone 1-800-SPRINGER, fax (201) 348-4505, e-mail orders-ny@springer-sbm.com, or visit www.springeronline.com. Apress Media, LLC is a California LLC and the sole member (owner) is Springer Science + Business Media Finance Inc (SSBM Finance Inc). SSBM Finance Inc is a **Delaware** corporation.

For information on translations, please e-mail booktranslations@springernature.com; for reprint, paperback, or audio rights, please e-mail bookpermissions@springernature.com.

Apress titles may be purchased in bulk for academic, corporate, or promotional use. eBook versions and licenses are also available for most titles. For more information, reference our Print and eBook Bulk Sales web page at http://www.apress.com/bulk-sales.

Any source code or other supplementary material referenced by the author in this book is available to readers on GitHub via the book's product page, located at www.apress.com/9781484274156. For more detailed information, please visit http://www.apress.com/source-code.

Printed on acid-free paper

For Edwina, Travis, Vivian, and Ruby

—RF

****

*To my father, who has always inspired me
to live a life of curiosity and exploration.*

—AS

****

For Sonia, Aastha, and Shaun

—KG

Contents

About the Authors vii

Acknowledgments ix

Preface. xi

Chapter 1:   Introduction 1

Chapter 2:   Initial Preparation 7

Chapter 3:   Operating Model – Governance, Sponsorship,
              and Framework 15

Chapter 4:   Opportunity Identification 29

Chapter 5:   Opportunity Assessment. 45

Chapter 6:   Solution Design 65

Chapter 7:   Solution Deployment, Maintenance, and Retirement 97

Chapter 8:   Organizational Structure 123

Chapter 9:   Development Methodologies and Framework. 137

Chapter 10: Planning for the Future 155

Chapter 11: Challenges and Pitfalls 165

Chapter 12: Summary. 173

Appendix 179

Index 201

About the Authors

Robert Fantina is an acknowledged process improvement expert and has worked closely with Andriy and Kamal in establishing an effective and successful RPA program at his most recent place of employment. He is the author of eight books, including *Practical Software Process Improvement* and *Your Customers' Perception of Quality: What It Means to Your Bottom Line and How to Control It* (co-author Baboo Kureemun). His paper "Successful Software Process Implementation" was published in the journal *Software Quality Professional*. He has given presentations on process improvement and quality at conferences in Atlanta, Georgia; Los Angeles, California; and Orlando, Florida, among other places.

Andriy Storozhuk has extensive expertise in all aspects of Lean Six Sigma, and he has utilized those skills in a variety of environments and methodologies. He has incorporated that knowledge, along with knowledge received in his university studies and constant ongoing training, in all of his work. Most recently, he brought that knowledge and experience to bear in creating an RPA program at his current place of employment.

Kamal Goyal has been working exclusively in the field of RPA for the last six years. He has been instrumental in establishing the required infrastructure at several companies, including where he currently works. He, too, is constantly learning to keep abreast of current trends within IT.

Acknowledgments

This book was created based on our experiences at several different companies in Canada and the United States. We are grateful for the opportunities afforded to us by these organizations.

We also wish to thank the various experts whose words we quoted occasionally in the book. Their work has been invaluable to us as we embarked, some time ago, on our own RPA (Robotic Process Automation) journey.

We are also very grateful to our editors, Mark Powers and Laura Berendson, for their guidance and assistance throughout the process of writing. We also want to acknowledge and thank Shivangi Ramanchandran for her feedback and for giving us the opportunity of working with her team.

Lastly, we would like to acknowledge our wives who we may have, of necessity, neglected at different times throughout the process and who maintained their constant support for us regardless.

Preface

The book in your hands represents decades of experience of the authors. We have worked, as employees and contractors, in many different industries, on very diverse projects, in companies that range from very progressive to those that shun even any discussion of change.

But in this millennium, organizations resist change at their existential peril. Many companies that were leaders in their fields at one point are now either insignificant or nonexistent. Think "BlackBerry" for one key company that took a major fall, reducing its employees from a high of nearly 17,000 to less than 4,000 today.

That is only one stunning example of many. But it is indicative of what happens when a company fails to keep up with important trends and doesn't make the changes necessary to remain competitive.

We have long recognized the need to keep our own skills current, to maintain our own marketability, and provide the best possible service to the companies that employ us. Our work in Robotic Process Automation is the latest trend we have seen and worked in and one we find exciting to the point of being revolutionary. We often hear of some "revolutionary" new tool or methodology, but RPA has proven itself time and time again; we have seen it in action.

As we discussed our experience in RPA, and together implemented an RPA program in a company within the financial services industry, our belief in it as an excellent tool to assist any company in any industry to remain profitable and competitive was strengthened. Yet our discussions and our reading also demonstrated that many organizations don't understand RPA and therefore don't recognize its benefits. But we know from experience how beneficial it can be.

The amount of money that companies spend on training for their employees in any given year can be significant. It's wonderful that organizations are willing to invest in their people, and we encourage them to continue to do so, as we also encourage employees to take full advantage of company training programs. But, too often, employees come back into the office after training, enthusiastic about what they have learned and anxious to implement it, but the structure for doing so simply isn't there. It could be that management likes the idea, but doesn't fully understand it and is hesitant to implement it, or other employees are simply resistant to change. After all, if someone has been doing their job successfully, why change the way it's being done?

This book answers that question, along with many others. Work may be done very effectively, but not efficiently. For example, an employee who processes paperwork all day long, looking up account numbers, copying them to an invoice, etc., may be very good at their job. But if the work can be automated, it's highly possible that a week's worth of invoices can be completed and sent to the customer in the amount of time it takes a person to process a single one.

These are the kinds of issues management must look for. How can more work be done with fewer resources? How can work be more accurate? Managers and leaders in companies we have worked for have struggled with these issues, and some of them have seen automation as one of the answers.

Robotic Process Automation is not mysterious, but it must be understood to be successful. There are some prerequisites to implementation, and if you miss them, you increase your risk of failure. Within these pages, we provide the tools to help ensure your success.

We recommend that you read the book start to finish and then use it as an ongoing reference guide. You will find useful tips based on our experience, along with all the templates you will require. We ask specific questions within each chapter for you to consider in the context of your own organization.

The book is designed and written for business leaders, and that can mean anyone who wants to take the initiative to improve the efficiency and effectiveness of processes within their jurisdiction.

We wish you every success as you learn about, and then implement, Robotic Process Automation.

Robert Fantina

Andriy Storozhuk

Kamal Goyal

Introduction

Overview

For decades, we have been hearing about the "fast pace" of technology, the "rapidly changing technology environment," and the need for companies to be on the "cutting edge" in order to remain competitive or even relevant. Different methodologies have come and gone, some like Six Sigma remaining popular and effective and others like the Capability Maturity Model (and its short-lived successor, the Capability Maturity Model Integrated) falling by the wayside after a respectable run.

All of these have served, or continue to serve, a purpose. These models have been adopted successfully by many organizations. But process improvement is an ongoing initiative. What can be done next in terms of current processes, here and now, that will increase efficiency, benefit customers, and improve profitability? One option is Robotic Process Automation (RPA).

Cristina-Claudia Osman, writing in *Informatica Economica*, described RPA as follows:

> *The emerging of new digital tools supports organizations to improve their business processes by increasing efficiency and agility, or reducing errors and costs. Robotic Process Automation (RPA) tools are designed to perform manual and repetitive tasks of human employees using trained robots. This is different than other traditional software because the robots communicate with other information systems by means of front-end. RPA helps companies to reduce employees' workload and errors, but also to save costs.*

To summarize, RPA means taking repetitive, manual processes and creating a "Bot" (robot) to perform them, thus freeing up the time of personnel who were doing these tasks to focus on more important issues, issues that often require human insight and decision-making, not just following routine and repetitive actions. These more complex tasks involving cognitive thought are often more directly customer facing. Additionally, automating processes reduces human error and costly rework. It can also improve morale by enabling employees to work on tasks that require thought, rather than simply repeating rote actions.

■ **Pause and Consider**   Automating certain processes enables people to focus on more complex processes and tasks and those that require decision-making and/or creativity. A Bot will do exactly what it is programmed to do, and that is most often jobs that don't require thought. Consider how this will help to gain buy-in for a new RPA initiative.

Benefits

The benefits of RPA are clear. As Timothy Driscoll, writing in *Strategic Finance*, said:

> Designed properly, software robots can operate 24/7 at a lower cost than humans while delivering higher-quality and scalable output. This set of benefits can make an appealing value-creating business case. RPA initiatives are likely to target processes that are labor- and transaction-intensive, where people are performing recurring tasks that can be redesigned as rules-based activities performed by robots or software tools. The benefits of RPA initiatives may include, but aren't limited to, the key measures in Table[1].

Efficiency/Cost Metrics	Effectiveness/Quality Metrics	Risk/Compliance Metrics
% Headcount/cost reduction	% Accuracy improvement	% Operating risk improvement
% Productivity increase	% Cycle-time reduction	% Increase in automated controls
% Daily throughput increase	% Client-satisfaction increase	

---

[1] https://sfmagazine.com/post-entry/march-2018-value-through-robotic-process-automation/.

As shown, RPA can achieve benefits that the management of any organization would welcome. We will now discuss briefly how these can be applied. Detail on how to achieve these and additional benefits is contained in subsequent chapters.

Practical Applications

In order to achieve these and additional benefits, RPA can be applied to a wide variety of processes. There are four key characteristics to look for when determining if a process can easily be automated:

1. Is it repetitive?

    - Do the same actions need to be performed over and over, for different clients, employees, etc.? For example, if an invoice is created manually, a clerk must obtain the name of the item or service purchased and the cost of that item or service; they must find the name, address, and account number of the purchaser and insert all the information onto an invoice template. They then place it in an envelope and send it to the mailroom for postage.

2. Would the cost of automating the process exceed the cost of continuing to do it manually?

    - While a manual process may be very repetitive, it may cost more to automate it than to continue doing it manually. On the surface, such a process would not be a good candidate for automation.

3. Is automation required for competitive value?

    - Saving money is not the only motivation for automating a process. Perhaps the cost of automating a process will exceed the dollar savings, but it should be automated to maintain a competitive advantage.

4. Would automating a process better enable the organization to fulfill government regulations?

    - Most industries adhere to various kinds of government regulations. Automating processes to ensure that required work is done and reports are issued on time can be highly beneficial.

In looking at any process automation potential, a "Yes" to question 1 is required in order to continue investigation. The repetitive nature of the process is key for automation. And while the cost of automation may exceed the financial benefits accrued by it, this does not eliminate the process from consideration for automation, since there may be competitive or regulatory benefits that go beyond immediate financial considerations.

Structure of the Book

How do you apply the principles to achieve the desired benefits? This book is structured to enable the novice to RPA to successfully implement an RPA program within their company. The first two chapters describe the groundwork required to implement RPA successfully. The remaining chapters describe the activities in specific phases of RPA implementation and how to monitor to ensure success.

Chapter 1. Overview. Introduction to RPA, what it is, how it is growing throughout IT organizations, and how it can benefit your organization.

Chapter 2. Initial Preparation. What is required to introduce a successful RPA program in any organization, larger or small? What are the key components that must be in place before you begin? How do you identify "early adopters" who will be key to your success? What are the hidden land mines, and how can you avoid them?

Chapter 3. Operating Model – Governance, Sponsorship, and Framework. How much structure is needed? How flexible should that structure be? What gates should exist in order to assure that approval for a project remains valid, even if circumstances change? Who are the key stakeholders who will be responsible for the RPA program? Why is cancelling a project in a later phase not a "failure"?

Chapter 4. Opportunity Identification. How do you solicit opportunities for automation? How do you best introduce the idea of automating repetitive manual processes? What forms will help you to best identify opportunities? Who (what roles) must be involved at this stage?

Chapter 5. Opportunity Assessment. Once you have identified some possible candidates for automation, how do you assess them? What should you look for? What documentation is required? What do you present to the Governance Committee? How do you handle ambivalence ("this might be a good opportunity if…")? Who (what roles) must be involved at this stage?

Chapter 6. Solution Design. Once approved, how do you design the solution? Who is involved? What documentation is required?

Chapter 7. Solution Deployment, Maintenance, and Retirement. Once designed, what is required pre-deployment? What steps must be taken to ensure the best possible chance for success? What documentation is required? Who (what roles) must be involved at this stage?

Chapter 8. Organizational Structure. What is the best structure for your new RPA program? You will need to know what currently exists in your organization. We detail the three most common structures: centralized, decentralized, and hub and spoke.

Chapter 9. Development Methodologies and Framework. Different organizations use different methodologies, such as a traditional Waterfall methodology, Agile, or a hybrid. Whatever methodology your organization uses, we describe the commonalities and how to best assure RPA success, regardless of the methodology.

Chapter 10. Planning for the Future. As technologies change, how will you adapt? Also, as processes change, what alterations may be required for existing Bots? How can you build in the ability to change, even when the future is unknown? This chapter provides the necessary guidance to assure that your new RPA program will continue to provide benefits far into the future.

Chapter 11. Challenges and Pitfalls. You will encounter many challenges as you implement an RPA program. We have detailed some of them in early chapters, but in this chapter, we will summarize them. What is the risk of half-hearted executive buy-in? How will you deal with line workers who feel their jobs may be jeopardized by automation (and how will you handle that situation if that is a real threat)? How do you work with a COE (Center of Excellence) most effectively, and on whom do those responsibilities fall if there is no COE in your organization?

Chapter 12. Summary. In this concluding chapter, we focus on some of the key highlights of advantages and why RPA may be ideal for your organization. We also briefly note some of the cautions to be aware of as you begin to introduce RPA.

Appendix. The Appendix contains all the documentation you need, as described within this book. This includes, but is not limited to, a request form (the basic starting point – eventually, this will be completed by the requestor, but initially, you will complete it with the requestor, since they will be unfamiliar with it), a risk assessment table, a feasibility assessment table, requirements documents (functional and nonfunctional) templates, and many others. These, of course, can and should be adapted to your particular needs, but they provide you with an important starting point. Often it is best, at first, to use them as they currently exist. Your ongoing experience with RPA will lead you to adapt them for greater effectiveness in your particular organization.

While this book will be an ongoing reference for any RPA initiative, it is recommended that it be read start to finish initially. This way, the person new to RPA will have an excellent overview and know where to look within the book for more detailed information when questions arise.

We wish you much success as you start your RPA journey!

Initial Preparation

Need for RPA

The introduction of an RPA initiative in any organization usually stems from one of three sources:

1. Senior management has learned of RPA and believes it will be beneficial to the organization. This may result from a review of a business journal that discussed RPA, a casual conversation with a peer from another organization, or any of a variety of sources.

2. A mid-level manager is aware of RPA and has recognized that there are many repetitive, manual processes within their area that could be automated for better efficiency, fewer errors, and decreased expense.

3. A technology leader recognizes the importance of RPA. Since RPA is greatly increasing in popularity, technology leaders are discussing it more frequently, and it is appearing regularly in a variety of journals.

© Robert Fantina, Andriy Storozhuk, Kamal Goyal 2022
R. Fantina et al., *Introducing Robotic Process Automation to Your Organization*,
https://doi.org/10.1007/978-1-4842-7416-3_2

The initial steps will be different depending on the reason that RPA is being introduced to your organization.

■ **Pause and Consider**   Why is RPA being introduced in your area? Is it reason 1, 2, or 3 mentioned earlier or a combination? Maybe a senior manager has heard of RPA and has asked for your opinion: Do you see it benefiting your department? If so, in what ways? Do you think it could be used throughout the company? Perhaps you have been asked to pilot it in your department.

---

    1.   Senior management decision.

Key to the success of an RPA implementation, regardless of why it is being introduced, is stakeholder buy-in. Has one senior manager decreed that RPA is to be introduced? If that's the case, other key members of the senior management team must buy into the concept before any investment in RPA (time, resources, infrastructure, etc.) begins. Additionally, and even more importantly, middle managers must see the value of it.

■ **Tip**   Do not neglect the importance of middle-management and staff buy-in. Resistance at this level can quickly and effectively sabotage an RPA initiative.

If additional members of the senior management team must be brought in, do your homework. There are several things you should present, at a fairly high level, that will assist you in convincing them of the benefits of RPA to your organization:

    I.   Industry standards: Many organizations are moving toward automating manual, routine tasks that require no thought. For example, in financial services, if every time a client makes a transaction, their financial adviser must be notified; this is a process that can be automated. The client name can be read from a database, matched to the financial adviser, their email address located, and then a template can be automatically completed and emailed to the adviser.

Do the necessary research to see how RPA is being implemented within your industry.

Pause and Consider In many organizations, the employees are working hard, focusing on the tasks at hand. They often don't see how their work could be automated; they may have been doing the same job the same way for years. They don't see how steps could be eliminated or the process improved. How will you help people look for automation possibilities?

II. Specific examples in your organization: What are some specific processes within your organization that could benefit from automation? You may need to do some research to discover these. This will involve working with other middle managers and their teams, which will also assist in getting buy-in from this level.

Pause and Consider For many people, automating tasks means layoffs. How will you alleviate this fear? Or, if it is a definite possibility, how will you handle it? Managers at all levels are loath to see their staff reduced in size. And their team members will resist any measures that may cost them their jobs. However, headcount reductions could be the goal or even the unintended outcome, and this must be addressed.

III. The opportunity to retrain and redeploy employees whose work is to be automated can also be a benefit.

Pause and Consider In many organizations, redeploying personnel to other responsibilities that require more thought than routine, manual tasks is one goal of automation.

One way of addressing this is an HR talent development strategy. Automation can free resources to retrain and move to other areas where the employee can use their skills beyond performing rote, manual tasks. Who will you need to speak to in your organization to accomplish this?

IV. Benefits: As you discuss automation within your industry, and some examples within your own organization, quantify and/or qualify the potential benefits. This could include, but not be limited to, any or all of the following:

- Customer satisfaction: Clients (internal or external) will receive more timely information.

- Improved accuracy: No more "human error" in an automated process.

- Reduced rework

- Financial savings: Reduced headcount, more efficient use of personnel.

- Competitive advantage: Being able to provide information either faster to clients than your competitors do, or at least keeping up with them.

- Time savings: Required information gets to decision-makers more quickly.

- Government regulations adhered to more promptly: Time-consuming reports can be automated and sent to the appropriate government bodies in a timely manner, reducing or eliminating the risk of fines.

- Increased employee morale, resulting in better job performance and reduced turnover.

Remember that the initially estimated benefits are just that: initial estimates. Do not overpromise. Point out the general trends and benefits in your industry, highlight some likely candidates for automation in your organization, and give a general estimate of potential benefits. *Be careful not to guarantee that certain benefits will accrue.* Doing so is a surefire way to have the entire program cancelled when those benefits do not materialize. At this point, you are giving estimates based on industry results and your own expertise. But RPA is new to your organization, and it will take time and practice to improve your estimating accuracy.

Tip You don't need to guarantee certain results to obtain buy-in; you can't at this point anyway. Explain the potential benefits, and recommend trying the new program with a few processes.

If senior management has mandated an RPA program, before you begin implementation, you must get middle-management buy-in. Use the information in step 2, below, to do so.

    2.   Mid-level management idea.

Some of the steps will be the same, regardless of the source of the idea. The main difference in this scenario is that you do not have a senior leader sponsoring the idea. That is a challenge, but not an insurmountable one.

Again, it will be necessary to do the required research on RPA overall and in your industry specifically. Look at both the gains and the drawbacks. Get the answers to these questions:

- What kinds of processes within your industry are being automated?
  - Why are they being automated?
  - What are the results?
- What are the challenges to successful RPA implementation within your industry?
  - How have they been overcome (if they have been)?
  - How could they be overcome in your organization? Who are the key people to speak to about them?

■ **Tip**   Some of the challenges have been mentioned earlier, including employee fear of layoffs.

Once you have thoroughly investigated RPA benefits and implementation within your industry, you need to identify allies. Who will be the "early adopters"? These are the people you will need to bring on board to support the initiative. (Note: if senior management has mandated RPA implementation, use these steps to get middle-management buy-in).

■ **Pause and Consider**   For early adopters (people who will be willing to start using RPA almost immediately), consider first your peers who are likely to be interested. Who manages areas where there are repetitive, manual processes? Who is frustrated by the lack of time their staff has to work with clients, because they must perform repetitive, manual processes? Identifying these people will be key to your success.

Informal conversations with peers will assist you in knowing who might be most interested. Once you have a few names, be direct. Explain to them that you are considering proposing RPA to senior management. Ask these questions:

- What processes do you have that are repetitive and manual that might be good candidates for automation?

- Would you be willing to pilot RPA with one of your processes?

- How would your team react to RPA? Would there be fear of layoffs, or would team members be anxious to spend less time on these manual processes and more on "thought-provoking tasks" (tasks that take actual thought, not just performing those rote tasks)?

- What time frame might be workable for them to try an automation?

■ **Pause and Consider**   Exactly how you will progress with RPA implementation is unknown at this stage, but you must still have a clear strategy. People are uncomfortable with uncertainty and will cling to what is known and familiar. How will you articulate a strategy that is not step-by-step specific, but that still outlines the overall plan?

■ **Tip**   While you don't need a crowd of middle managers to buy-in at this point, getting at least a few will strengthen your argument with senior management.

Compile the information you have received. If you don't feel confident that you have the necessary support at the middle-management level, do not proceed! Continue to work to build that support, but moving forward without some enthusiastic support is a recipe for disaster.

Assuming you receive the support of a sufficient number of middle managers (what that number is will differ depending on the size of the company), prepare a short presentation for senior management. It should include the following information:

- Overview of RPA

  - What it is

- Review of RPA successes within your industry

- Review of RPA challenges within your industry
  - Discussion of how you anticipate overcoming those challenges
- List of middle-managers willing to pilot the program
  - List of potential processes to be automated

■ **Pause and Consider**   What is the correct amount of information for senior management? Remember, senior management members don't need all the detailed information to make a decision. They need to understand at a high level what you want to do, how you will do it, and how it will impact them. Keep these things in mind as you prepare and deliver your presentation. After you create your presentation, be willing to edit it, to remove details that may be required for middle management, but not for senior management.

Be prepared to answer questions, but don't hesitate to admit if you need to further research something that has been asked. However, if your presentation is at the right level (see "Pause and Consider"), there should be few questions. But if there are questions, get the answers and email the senior management team with them as soon as possible. If in getting the answers, any of your earlier assumptions or information change, be sure to relay that information.

■ **Tip**   If you can't get buy-in right now, do some more research. Learn what your peers in the industry are doing and understand both their successes and failures. Do not attempt to implement a new program without the required buy-in. Doing so may result in failure.

Once you have received senior management's approval, which will only occur after you have obtained the support of some members of middle management, you are ready to begin.

Remember, regardless of the trigger for introducing RPA implementation in your organization, it is vital that you obtain senior management and middle-management buy-in.

3.   A technology leader recognizes the importance of RPA.

The same steps in number 2 will be used: find early adopters, identify potential projects, and make a convincing presentation to senior management. However, this could be more challenging. Senior management members, often more business-focused than technology-focused, may see this as some "cool new thing" that technology is excited about. You will need to assure that the benefits to the business are highlighted.

There are many reasons for the introduction of RPA into an organization: it could be because a senior manager has become familiar with the concept and sees the potential advantages to the organization. Perhaps a mid-level manager, dealing with many repetitive, manual processes, recognizes that automation will be highly beneficial. Or a technology leader could see how their skills could be utilized to increase efficiency. The way you introduce RPA to your organization will depend on why it is being introduced. The overall benefit is the same: increased automation of previously manually performed tasks. But the successful introduction of RPA requires that you know how best to present it, and knowing why it is being brought in is key to determining how to begin it.

Once you have accomplished that, you are ready to proceed to the next step.

■ **Tip** In Chapter 4, we introduce a case study and demonstrate in that and through succeeding chapters how each phase of RPA would apply to it. You may find reviewing that case study helpful as you work with senior and middle management.

Operating Model – Governance, Sponsorship, and Framework

Governance, sponsorship, and framework will be major factors in success. We will describe what roles within a typical organization should be involved and what their responsibilities will be.

James Chen, writing in *Investopedia*, describes **governance** in this way: "Corporate governance is the system of rules, practices, and processes by which a firm is directed and controlled. Corporate governance essentially

© Robert Fantina, Andriy Storozhuk, Kamal Goyal 2022
R. Fantina et al., *Introducing Robotic Process Automation to Your Organization*,
https://doi.org/10.1007/978-1-4842-7416-3_3

involves balancing the interests of a company's many stakeholders, such as shareholders, senior management executives, customers, suppliers, financiers, the government, and the community."[1] For RPA, governance is not quite that wide ranging, and we will describe what it needs to accomplish for RPA success.

The website Project Smart describes **project sponsorship** this way: "Project sponsorship is an active senior management role, responsible for identifying the business need, problem or opportunity. The sponsor ensures the project remains a viable proposition and that benefits are realised, resolving any issues outside the control of the project manager."[2] The sponsor is usually the person responsible for the budget of the area where an RPA project is requested.

"A **framework** is a loose but incomplete structure which leaves room for other practices and tools to be included but provides much of the process required."[3]

The relationship between the three concepts is clearly seen. Governance consists of the rules and processes, sponsorship is the authority required to ensure their survival and evolution, and they are all performed within the framework.

■ **Pause and Consider**   Is there effective governance and sponsorship within your organization today? Have you seen either or both being effective either in your current company or a previous one? It is possible (perhaps even likely) that you have not. Do not underestimate the importance of good governance and effective sponsorship when beginning your RPA implementation journey. Doing so will doom the project from the start.

- Governance

The RPA Solution Lifecycle includes the following; bear in mind that there may be some differences depending on the individual needs of specific organizations.

- Opportunity identification (detailed in Chapter 4)

---

[1] www.investopedia.com/terms/c/corporategovernance.asp.
[2] www.projectsmart.co.uk/project-sponsorship.php
[3] https://pm.stackexchange.com/questions/3791/what-is-the-difference-between-framework-vs-methodology#:~:text=A%20methodology%20is%20a%20set,much%20of%20the%20process%20required.

1. Request form

    This is the basic request that can occur in different ways; it could be an informal hallway conversation, followed by a meeting to capture required details to move forward with the request, or a requestor may have submitted a request form (see Appendix 1) which provides all the required information.

    The information on the request form represents the first look at the opportunity, and consequently, there is much unknown information at this point.

    Typically, at this point, the following information is captured:

    - High-level description of the process
    - Impacted business areas (upstream and downstream processes) and systems or tools
    - Business drivers and criticality
    - Estimation of savings (FTE hours, etc.) and benefits (improved quality, improved service, etc.)

■ **Tip**   While initial requests may be in the form of emails, or even informal hallway conversations, standardizing the request form and encouraging its use will be highly beneficial.

There may, however, be situations where information provided on the request form is sufficient to conclude that automation is not feasible under current circumstances, possibly because of the process complexity, RPA-limited capability, or even the potentially short life span of the solution. The requestor should be advised of this and given any possible recommendations to assist in overcoming whatever issues were to be solved by automation.

Once an initial request has been submitted that didn't rise any feasibility concerns, it is ready to move to the next step.

■ **Tip**   Remember, this is the initial look at the request; not a lot of detail is yet available. If no feasibility concerns are raised here, and the request is given approval to proceed, that does not mean that such concerns may not be uncovered in future phases, causing cancellation of the project. If that happens, it must not be considered a failure. It means that the iterative process has worked as it should.

---

2. Process discovery

   At this stage, the referred opportunity is assessed in detail to capture the following:

   - End-to-end (E2E) process steps

   - Roles and users involved in the process

   - List of systems and tools used

   - Volumes (hourly, daily, weekly, etc.)

   - Process time for each step and other process performance metrics that would help accurately estimate savings and benefits

   - Process constraints

   - Proposed process steps to be automated

   - Any other information that was triggered by the request form

   All this information is required to move to the next step, opportunity assessment, which will estimate RPA value and RPA feasibility as a solution.

- Opportunity Assessment (detailed in Chapter 5)

Several things are accomplished in this phase:

1. Understand the current process flow

   - The process flow was examined at a high level during Opportunity Identification. Now it is detailed, and the inputs, activities, outputs, and roles responsible are documented.

2. Impact analysis

   - Why should this be done? During intake, some initial estimates may have been made regarding time saved, compliance adhered to, etc. Now these are studied for more accuracy.

   - What is the investment of time and resources to create a Bot to automate the process? Is the ROI (return on investment) sufficient to justify the expenditure? While this is a decision for the steering committee to make, the RPA team will make a recommendation.

3. Suitability analysis

   • Here we must evaluate if the requested process is a good candidate for automation. Is it sufficiently repetitive? Are there few decisions that need to be made that can be referred for manual handling? These and other questions will be answered to determine suitability.

4. Risk analysis

   • What are the risks involved in automating the process? What are the risks involved in *not* automating it?

   • Solution feasibility decision process

All compiled information is presented to the decision-makers to evaluate and to ensure that the RPA solution will bring sufficient ROI, and expected benefits are aligned with the enterprise-wide strategy.

After the decision is made to proceed with RPA solution development, the opportunity moves to the solution design phase.

   • Solution Design phase (detailed in Chapter 6)

   1. Plan

   2. Build

   3. Test

      • Each phase is performed according to company standards, using "Waterfall" methodology, Agile, or whatever other methodology is generally used.

After successful testing, and passing business acceptance, the RPA solution is released to production entering operational phase.

   • Solution deployment and maintenance phase (detailed in Chapter 7)

   1. Release

   2. Operate

   3. Monitor

   • Solution retirement

   1. Sunset (end-of-life) strategies or plan

The information provided previously is shown in process-flow format in Figure 3-1.

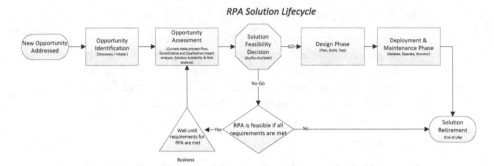

Figure 3-1. RPA Solution Lifecycle

The components of effective governance may seem complex, but the need for this structure has been demonstrated time and again. Identifying and assessing the opportunities, and having repeatable processes and tools for doing so, will go a long way to ensuring the success of your new RPA initiative.

- Sponsorship

The RPA sponsor will be responsible not for performing these tasks, but for assuring that the outcomes or results are aligned with, and are adding value to, enterprise-wide strategic direction. He/she establishes program vision, objectives, and goals, removes any impediments, and ensures that appropriate resources are allocated necessary to the program's success. Typically, the RPA sponsor is a member of the executive team, who oversees the area where the RPA program is being introduced. It is fine either for business or IT to own the program. However, it is important to remember that the RPA team needs to be cross-functional and should include the following:

- – Business
- – IT
- – Security
- – Risk and compliance
- – Finance
- – HR
- – Marketing (when appropriate)

Stakeholders/subject matter experts from these areas must be involved in creating a collaborative environment since, as is mentioned in the article, "A Blueprint for Implementing Robotic Process Automation Successfully,"[4] RPA

---

[4]https://pt.slideshare.net/CFB_Bots/a-blueprint-for-implementing-robotic-process-automation-successfully/11. Accessed on March 16, 2021.

has the potential to become a disruptive change agent for true and impactful digital transformation.

The RPA champion, on the other hand, will be responsible for promoting RPA technologies to address business challenges and is accountable for RPA adoption across the business areas where the program is introduced. Ideally, the RPA champion is one who has a deep knowledge of business operations and has the necessary influence to promote and overcome resistance to RPA adoption by providing timely resolution to any roadblocks.[5] It is optimal to have several RPA champions representing different business areas to maximize return value of the RPA program and ensure enterprise-wide adoption.

■ **Pause and Consider** Sponsors and champions should have passion and vested interests in the success of the RPA program. Have you done the necessary work to assure that the people you want as sponsors or champions believe in RPA? Do not proceed with people who are either only required by their manager/leader to participate or who you know are not convinced of the potential of RPA.

Ensuring that there is a senior manager who is invested in, and responsible for, the success of the RPA program is vital to success. They should have a compelling reason for assuring its success.

Framework

When we refer to framework, we mean a general overview of how the program is structured. This is not written in stone; it will differ from company to company and even between different business units within the same company. But there are some basic foundational pieces.

As we mentioned before, it is critical that RPA program teams are cross-functional and multidisciplinary and should include representatives from all impacted areas within business operations and control support functions. A fully functional core RPA delivery team should consist of the following roles:

- Process analyst/designer
- RPA solution architect
- RPA developer
- RPA management and support

Process Analyst/Designer

This role requires understanding the process that needs to be automated, checking the feasibility of automating that particular process and, if it makes

---

[5] https://medium.com/@cfb_bots/a-blueprint-for-implementing-robotic-process-automation-successfully-88e5f7ff7400.

sense to do so, designing and developing the future state of that process. The process analyst/designer is experienced with business processes and is able to define the vision of the desired processes. Since there might be one or more people in that role, it is very important to remember that the process analyst/designer has to work closely with people who perform the current process in order to accurately capture its details. The process analyst/designer must understand and evaluate these details in order to assess if the process is a good fit for RPA and that the deployed solution (Bot) won't expose operations to any risks and will yield sufficient benefits. To do so, the process analyst/designer will also be familiar with the capability and limitations of RPA tools and be able to provide basic overview training to the process owners. The process analyst/designer is the driver of the specification for the RPA solution, at the same time acting as an owner of any other related documentation such as feasibility, impact and risk assessments as well as business acceptance handoffs, etc. In addition, they should be responsible for incorporating back into the specification any feedback and changes made during the development and testing phase.

RPA Solution Architect

RPA solution architects traditionally come from software development or automation backgrounds. The RPA arena is growing very fast and is assisted by simple Task Bots that execute repetitive, rule-based tasks. These are easy to build to a new generation of IQ Bots that allow the addition of cognitive processing capabilities, enabling the process to deal with unstructured data. Also, there are plenty of packaged enterprise software offerings, such as AutomationAnywhere, BluePrism, UIPath, etc. The RPA solution architects need to be familiar with not only the RPA capabilities available today but must also be knowledgeable about the following:

- Automation strategies
- Digital solution delivery models and frameworks (Agile, DAD, DevOps, etc.)
- IT infrastructure and business operations
- Products or services that will be addressed in an RPA program

This expertise is required to assist and guide the organization throughout the RPA journey in building sustainable, effective (impactful), and scalable solutions.

RPA Developer

The main responsibilities of the RPA developer are to design, develop, and maintain the automation solutions using the RPA tooling. Depending upon the tooling used, this role will require some developer-level expertise. If you're looking to automate a process that involves rules and logic such as loops, if/

else statements, variables, and exceptions, it requires programming, and the RPA developer should have the required experience in that discipline. Keep in mind there are multiple ways to automate a process; therefore, good technical knowledge of automation tools is a huge asset.

That the RPA developer is a part of the cross-functional design team means they are working closely with the RPA solution architect and the process analyst/designer to find the most effective and efficient way to develop, test, and, after successful acceptance by the process owner, deploy to production the automation solution. It is important to remember that the released-to-production solution should be accompanied by supporting documentation to help developers and operations personnel understand the logic and content of the automated solution.

RPA Management and Support

Once the RPA solutions (Bots) have been deployed, there will need to be allocated resources who will monitor their execution and handle errors, fixes, and small enhancements in a timely manner. These resources will also manage the process execution schedule. In addition, the RPA management and support team will be accountable to maintain and advance overall the RPA ecosystem by monitoring and ensuring system health, maximizing runtime license utilization, introducing new automation capabilities, etc. Depending on the number of Bots deployed and the complexity of the RPA ecosystem, more than one person would be responsible for handling some of the above-mentioned functions. Also, it is beneficial to break down those high-level functions into smaller chunks in order to avoid overcomplicating things. Remember, the main goal of the tactical support is to maintain speed and agility and not to create excessive layers of bureaucracy.

Pause and Consider   The titles may differ from organization to organization. However, the titles are unimportant; the role descriptions discussed earlier are key. Who in your organization could perform these tasks? Are they familiar with RPA? Do they see value in it?

Before starting on an RPA path, an organization should decide if RPA initiatives will be led by in-house resources or be outsourced to an external third-party RPA provider. Also, an organization can choose a hybrid model that establishes a collaborative framework with an external provider and in-house resources. For instance, an external partner could focus on identification, development, and deployment of the solutions, and the in-house team will provide support and maintenance.

Before choosing a suitable path, an organization must be clear on a few aspects:

- Are there enough processes for RPA to justify investments (ROI)?

- Is the organization committed to long-term investment in an RPA program?

- Does the organization have the in-house skills and capability to start an RPA program?

- Is RPA a part of the long-term, meaning five- to ten-year, strategy of digital transformation of its operations and IT infrastructure or ecosystem?

Choosing the right path for an RPA program will make a huge difference between a successful RPA program and failure that leads to inadequate results and high cost. In all likelihood, this would cause withdrawal of the executive sponsorship and ultimately cut off financial investment in RPA.

■ **Tip**   Do not start an RPA initiative if these conditions are not met. Without these conditions, an RPA program has a much greater possibility of failure than success.

The RPA program's organizational model provides the structural framework for deploying RPA at the enterprise level. There are three models to consider[6]:

- Centralized

    This is typically designed as a Center of Excellence (COE; see Chapter 7 for more information about the COE) that is established within one executive office (business area) and manages the entire life cycle of the RPA program from developing standards, delivering RPA solutions from ideation to deployment, maintaining RPA infrastructure, and ensuring that adequate controls, risk management, and compliance are in place.

- Hub and spoke or federated

    The COE is established within one executive office and typically serves as a hub for decentralized spokes. In this case, the COE provides standard-setting, develops policy, provides training, and focuses on innovation and RPA program strategy. Spokes – RPA delivery teams – are

---

[6] The following information is adapted from this source: https://digital.gov/pdf/rpa-playbook.pdf.

primarily focused on developing and deploying RPA solutions to the business area they are assigned to or serve. This model requires close collaboration with the hub (COE) and spokes (RPA delivery teams).

- Decentralized

    Multiple individual RPA programs are established within an enterprise, with individual RPA frameworks operating under different executive branches or business units. This can be viewed as multiple sub-COE offices that are operating autonomously.

The optimal organizational model will depend on an individual RPA program strategy, size of the serving population, business complexity, management culture, organization design, risk tolerance, and many other factors. In the following chapters, we will get more into details on benefits and suitability factors and approaches for implementing an effective operating model for your enterprise's RPA program. Also, we will discuss how those models may change as the program becomes more mature.

Pause and Consider Which style seems right for your organization? How will you determine which is best? Does your organization currently have a Center of Excellence? If so, how does it operate? Who do you need to speak with from the COE about RPA? If there is no COE in your organization, is there an interest in starting one?

There are multiple product development methodologies for digital products to choose from. Those methods define necessary steps, the decision-making process, communication, and actions needed to be executed by different stakeholders in order to successfully deliver a digital product – the RPA solution.

The most common are:

- Waterfall

    This is a sequential framework where processes are executed in a liner manner and a product moves from one phase to another in a predefined or planned sequence. This method allows for high quality of product design, but its rigid structure does not allow a high degree of flexibility to change, and mistakes made at an earlier stage and discovered later can be very costly to resolve.

- Agile

  This methodology is widely adopted by software development companies, especially among start-ups and is centered around the idea of iterative development and delivery. This methodology allows teams to develop and release product features, enhancements, or changes frequently but in small segments. This allows for speed of value delivery to the customer and high adaptability or flexibility to changes in customer needs, and enables the quick identification of errors and mistakes.

- Waterfall-Agile hybrid

  This method is initially sequential and then turns into an iterative process through the development and delivery phases. This hybrid model creates a disciplined approach that allows the thorough investigation and understanding of customer needs and product specs prior to starting development and at the same time brings enough flexibility to iterate and adjust at the development and deployment phases.

- Lean

  This methodology relies on simple ideas like delivering the value defined from the customer's perspective. It uses concepts of Lean originally established in the manufacturing industry. In software development, it defines the minimum viable product first and then, through iterative cycles, continuously increases its value either through adding new features or improving its quality.

There is no concrete winner in this contest; it all depends on many factors, similar to those discussed earlier regarding selection of the appropriate operating model. It's very important to understand the environment you are in and the benefits and potential pitfalls of each methodology, within that context.

A well-structured framework is critical to the success of an RPA program. It can be modified over time as needs change or as deficiencies in the framework are identified, but starting with a solid framework to operate within will greatly contribute to success.

▧ **Pause and Consider**   What might the framework look like in your area? What are some of the basic structural requirements (roles, etc.)? What roles could potentially serve on the governance team? What roles would be appropriate for the decision-making body regarding whether or not to move ahead with requests? What documentation should be included? How will reporting be done? These and other considerations must be answered as you embark on an RPA program.

A flexible framework, with the required roles and processes included, is key to assuring that successes within the RPA initiative are not "one ofs," but are repeatable.

Governance, the rules, practices, and processes that will be used to control and monitor the RPA initiative will be key to success. Having effective **project sponsorship** – senior management responsible for assuring that RPA has the necessary support and achieves the anticipated benefits – is also a "must have" for RPA success. Assuring that there is a broad **framework** enables the people working within the RPA initiative to have a flexible structure to work within, assuring that they know what's expected, but still have the ability to adapt to changing needs and situations.

Opportunity Identification

Robotic Process Automation (RPA) is rapidly gaining popularity. Yet one of the main issues regards knowing which process to automate.

One of the most critical choices made with an RPA program is which processes are candidates for applying RPA.

While organizations are well aware of RPA, they are still trying to figure out the magic formula of RPA implementation. There isn't one, but the information in this book will assist in successfully implementing an RPA program at your company.

An article in *Sagacity Software* expressed this dilemma succinctly:

> The company that is implementing RPA technology must ensure that it is choosing the right processes to be modified. It is quite often the case that companies cannot automate all of their processes. Selecting the right processes can be a major challenge. If an organization chooses to implement the technology in too many of its processes, then it could cause a lot of disruption. At the same time, not implementing RPA in an adequate number of processes could hamper its effectiveness. Therefore, companies must find the right balance when it comes to implementation.[1]

---

[1] https://sagacitysoftware.co.in/what-are-the-challenges-of-rpa-implementation/.

Setting up the initial business case and identifying the right processes for automation can be complex. Attempting automation of unsuitable processes is one of the main causes of an organization's failure to realize the full potential of RPA capability to reimagine their business operations and propel toward digital transformation.

■ **Pause and Consider**   What challenges to RPA do you see in your organization? How will you address them? How will you manage expectations so that RPA is not seen as a "silver bullet" to resolve all problems, but will be viewed as an important step in saving time and reducing errors?

As we mentioned in a previous chapter, the RPA Solution Lifecycle starts with how RPA candidate processes are to be identified.

Please note that once you have established the RPA program at your organization, there will be two other related events that will require you to take the steps that follow in this and subsequent chapters: enhancements and defects. Enhancements occur when it is requested that a Bot that is handling a process is improved or changed to be able to perform more of the manual tasks. Also, there are times when a second process that is similar to the first is proposed for automation, and enhancing the existing Bot may be all that is required. In either of these cases, the request will be identified, assessed, etc.

Regarding defects, minor ones will be corrected without going through the process, but major ones should follow the process to assure that they are reasonable in terms of time and money to do the fix.

The basic request can occur in different ways: an informal hallway conversation followed by a meeting to capture required details to move forward with the request, or a requestor may have submitted a request form which provided required information.

The information on the request form represents the first look at the opportunity, and consequently, there is much unknown information at this point.

A request form provides the bare essentials of the request. It is used to decide whether or not to proceed to the next step or if the project should go no further at this time.

At this point, what is required is not sufficient information to decide on completing the project, but sufficient information to make one of the decisions referenced previously. If it is approved to move to the next step, more information will be gathered, and upon completion of that step, another determination will be made regarding moving forward.

A project can be cancelled at any time; there are many valid reasons for initially approving a project and then later cancelling it. For example, the preliminary estimates of dollar savings may, when more information is gathered in a later stage, prove to have been inflated. A more accurate estimate may preclude the benefits of spending money to automate the process. Or an automation that originally seemed to be fairly simple may, on closer look at a later stage, prove to be far more complex. As new information is gained, the sponsors must decide if there is still a sufficient return on investment (ROI) to proceed with the automation. If not, it is cancelled, and the stakeholders are advised of that fact and of the reasons.

A request that is accepted originally and later cancelled does not represent a failure. It is far better to expend some resources (time, money, etc.) on a project that appears feasible, but then, at some point during the project, cut losses and cancel it, than it is to proceed with the project to its conclusion and then realize it doesn't satisfy the need, went way over budget, etc.

Also, just because a request is either refused or cancelled after starting does not mean it will never be done. The requestor should be advised that due to current priorities and the fact that their request doesn't satisfy some criteria (e.g., insufficient savings, high complexity, etc.), it will not be a high-priority item at the present time, but will be re-evaluated in the future.

Request Form

This is a sample request form; you can also find it in Appendix 1.

RPA Opportunity – Request Form

Initiative Name			
Sponsor			
Champion			
How Does the Process Currently Work?			
What Is the Goal?			
Common Business Drivers/Business Criticality			
Impacted Business Areas		**Impacted Systems/Tools**	
Opportunity Type	**Duration of the Solution?**	**Estimated Operational Benefits (FTE Hours)**	
Priority		**Desired Delivery Date**	
Form Completed Date:		**Prepared by:**	

The typical components of a request form are the following:

- Name of requested project

- Sponsor: This is the person, usually an executive leader, within whose budget the organization requesting the automation resides.

- Champion: This is the leader directly responsible for the process that is proposed to be automated. They will also be responsible for removing any roadblocks that may occur and allocating required resources that would be needed during the automation development, testing, and deployment phases.

- Current process description: The current-state process is summarized in sufficient detail that anyone reading it has a comprehensive idea of the purpose, process flow, repetitive nature of the process, and the reason it is a candidate for automation.

- Implementation goals: This indicates what portion of the process, or the entire process, is proposed for automation. It briefly lists the benefits of automating the process.

- Business criticality: This section describes why this process must or should be automated at this time and how it is aligned to strategic business goals. Is there a competitive need? Is there exposure to high risks? Are clients complaining about something that would be resolved with this automation? Etc.

- Common business drivers: Some possible examples include, but are not limited to, the following.

  - Reduced cycle time

  - Improved consistency and quality of work

  - Allows staff to work on higher value activities

  - Etc.

- Impacted business areas: This lists the business units that would be impacted by this automation.

- Impacted systems/tools: There may be specific systems that have been purchased or developed that will be impacted; these should be listed here. Also, any tools, including Excel, Outlook, Word, etc., are listed here.

- Opportunity type: Note here if this is a one-time auto-mation, possibly to clean up months or years of backlog, or if it will be ongoing.

- Life span estimation: If the process is expected to be phased out within a certain time frame, or if the antici-pated purchase of a new tool in, for example, 3Q2X, will resolve it, note it here.

- Desired delivery date: Indicate when the automation is optimally desired.

- Estimated operational benefits: This is generally the sav-ings in time and cost, but not every requested automa-tion has a significant time and cost savings. Some are required to meet regulatory or competitive require-ments. Some are needed to reduce the risk of human error. If the automation will result in some time and cost savings or revenue gains, this should be explained in detail here.

  For example:

  - 5 staff members spend 3 hours a week on this process

  - $5 \times 3 = 15$ hours per week

  - 15 hours per week $\times$ 52 weeks = 780 hours per year

If a process is to be automated to save time and money, an annual savings of about 1,500 labor hours, or one full-time position, may be required to justify the solution development and maintenance cost. However, this will be different in different organizations, so it is important to establish the right criteria for yours.

- Form complete date: Self-explanatory.

- Prepared by: Self-explanatory.

■ **Tip** While initial requests may be in the form of emails, or even informal hallway conversations, standardizing the request form and encouraging its use will be highly beneficial.

■ **Pause and Consider**   Based on this information, what might a request form look like in your organization? Is there additional information that should be captured? The information above, and the template in Appendix 1, are what we have found useful as a basis, but different organizations may require slightly different components. However, the information shown here includes the basics.

As mentioned, one of the possible outcomes of this phase may be the rejection of the request. There may be situations where information provided on the request form is sufficient to conclude that automation is not feasible under current circumstances, possibly because of the process complexity, RPA-limited capability, or even short life span of the solution. The requestor should be advised of this and given any possible recommendations to assist in overcoming whatever issues were to be solved by automation.

Once an initial request has been submitted and didn't raise any feasibility concerns, it is ready to move to the next step.

Process Discovery Phase

It is vital to evaluate each individual process candidate to ensure RPA is a right solution to the current situation. It also ensures that the automation initiatives will be successful in your organization. In order for the RPA team and Governance Committee to make an informed decision about whether or not to proceed with automation development, they have to have sufficient information about the process candidate. The process discovery phase is preliminary to the opportunity assessment phase (please see Chapter 5), and it's where all information for making critical decisions is gathered. It's important to do it right in order to avoid beginning to automate a process that is not a suitable candidate for automation.

Process discovery can be done by going through the process-related documentation and extracting required information to provide a summary for the decision-makers. This approach is simple and quick; however, it can leave room for errors due to inaccurate assumptions or outdated information in the reference documentation. And what if there is no proper process documentation available?

As noted in BPM Resource Center:

> In many organizations, day to day activities may be so ingrained and routine that it's not always clear what the exact steps are in a given process. Different members of the organization may not know what happens to an item once it

leaves their hands. *Official process write-ups that you might find in the corporate operations or policy guides might not map what actually is being performed.*[2]

All of this leads to a high probability that when the solution design phase begins, the RPA team would discover extreme complexity, operational risks, and/or overstated benefits of what the automation of the process might yield.

Therefore, the recommended approach for process discovery can be as simple as gathering teams together (physically or virtually) and discussing and documenting (mapping and including a corresponding written description) how a particular process is actually performed. Through discussion and interviews with the people directly involved, you can usually discover the end-to-end steps of a process across teams, departments, divisions, etc. During this collaboration, you can also extract key metrics (volumes, costs, and other measurements) that can help determine if making improvement investments will yield the results you seek to meet strategic goals.

There are a number of BPMS (Business Process Management Software) tools that can facilitate or even partially automate the process discovery phase. By mining existing information systems and analyzing event and transaction logs, you can document the actual work being done (who did what, when, for how long). Today, new emerging technology tools can read in data from various systems and present that information in flow diagrams, charts, and in other ways to help you understand process steps, bottlenecks, and more.

However, often such tools are not available, and that is why identifying the right subject matter experts (SMEs) and meeting with them is key to success.

■ **Tip** Do not be discouraged if automation tools (event and transaction logs, etc.) are not readily available. Processes for automation can be readily identified through meetings with SMEs. Automation tools are an advantage, but not absolutely required.

As every organization suffers from some inherent inefficiencies in their processes, they require a method like process discovery that can help them access task-level data and the various nuances of interactions between humans and systems. While all companies establish standard operating procedures, in reality, they often go unheeded. Until these procedures are completely locked down, many employees will invent their own ways to operate. This is why the existing documentation of any process may not be reliable; it is necessary to know how the work is actually being done, which is often very different from

---

[2] www.what-is-bpm.com/bpm_primer/what_is_dynamic_business_process_management__bpm__.html.

how it is supposed to be done (according to existing documentation). In addition, many companies outsource their processes to other BPM vendors, leaving their leaders unfamiliar with the execution of their own processes.

In most cases, especially during the initial introduction of RPA, business processes are documented through manual process mapping. This consists of a high-level flow chart (map, usually using Visio or a similar software) and then describing each task in more detail in a separate Word document.

These two documents are enhanced by including information, as indicated earlier, regarding systems and tools, volume, time the process steps take, potential constraints to automation (if known at this time), and any other pertinent details as noted in the request form.

This information is then compiled for presentation to the governance committed.

■ **Pause and Consider**   As a program matures, the opportunity identification, intake, and process discovery should evolve from the optimal design of RPA solutions to optimal redesign of business processes. A robust slate of process improvement capabilities allows the high-performing RPA program to solve organization-wide business challenges and attain fully scaled, transformative impacts on operations performance and customer experience.[3]

Remember, a process map is used for designers, developers, Governance Committee members, and other stakeholders to understand the steps of the process. It is a high-level view of the inputs, outputs, and process steps in between. It assists in understanding what steps are most eligible for automation and how long each step takes to perform manually, so the time/cost savings can be determined.

There are several methods for creating the process map. You may rely on subject matter expertise. For this method, schedule time with people who actually perform the process. We cannot stress enough that often, there is a way the process is supposed to work and the way it actually works. It is vital to know how it actually works; what steps do people do to perform the process? This is what will be automated.

Map the process on a white board or on a tool like Visio or iGraphix.

First, document the steps, for example: "Start," "Find Account Number," "Validate Account Number," "Change Client Address," "Notify Client," "Notify Advisor," "End."

---

[3] http://microsoftrpa.com/page/2/#:~:text=A%20robust%20slate%20of%20 process,impact%20on%20operations%20and%20performance.

Then for each step, ask how long it currently takes. For example, finding the account number may mean searching a database and then copying the number so it can be validated against another database. How long does that take? Thirty seconds? Two minutes? Depending on the task, there could be a wide discrepancy in how long it takes to complete. In those cases, include a range (e.g., "Step 5: 30 seconds to 2 minutes").

Mark how long each task takes under the task on the map.

Another excellent way to map the process is to **video it**. For this method, it is necessary to schedule time with someone who actually does the process to video the tasks they perform in the process while they are performing them. As they are clicking on applications and drop-downs, or typing in information, they should walk you through the process in detail explaining what they are doing and why. For example, "in this text box marked 'description,' I copy the information from the body of the email that triggered this request and paste it into that box."

Once you have completed this, you can review it and create a process map from it.

Tip   If the project moves to the design phase (see Chapter 6), this video will be very helpful in documenting functional requirements.

If for some reason you cannot video the process, sitting with the person who performs it is another way to document the process (please note that all this can be done remotely). Note each screen and the actions that are performed on it, and then map the process as described earlier.

Once the first draft is created, it is beneficial to present it to a group of subject matter experts, the people who actually execute that process to make sure it reflects reality. Note that it is normal to discover that some people may tailor the process, either in terms of execution order or inserting an extra step or eliminating a step. Assure that there are no missing critical steps and that, if following the process as mapped, we would get desired results.

After the map has been created and validated, it is beneficial to provide some descriptive and more detailed information in a separate document. Remember that in any process map, each step has just a few words: "Identify Account Number," "Gather Requirements," etc. The detailed explanation helps the reader to better understand the process.

For each step in the process, the following information is provided in table format:

- — Input: What triggers the step? For all but the first step, the input is usually the output from the previous step.

- — Activity: What is actually done in this step? This is where information about accessing a particular system would be included.

- — Output: What is the result of the step? Using the previously referenced example, it could be "Account Number has been located."

- — Role(s): Who generally performs this activity? This is not a place for names, but for the role. Examples include, but are certainly not limited to, "Customer Service Rep," "Business Analyst," "New Business Manager," etc.

- — Metrics: Are there any numbers that are, or should be, collected with this step? Examples might include length of time to locate account number, time from initial request to response to customer, etc.

- — Notes: Include here any pertinent information that was not documented in the other sections of the step. This might include systems, applications or tools used, an explanation of possible exceptions, anticipated volumes, etc.

■ **Tip** Keep in mind that although this document provides additional information about the map, this is still an early stage document, so excessive detail is not required.

There are some circumstances under which this document is not required. For very simple and straightforward processes, the map may be sufficient. The inputs and outputs are clear, and the time each step takes is indicated. Also, if, during the initial conversation about the request, it appears that this request is not a good candidate for RPA, there is no point in spending time creating this document.

To summarize, at this stage, the referred opportunity has been assessed in detail, and you now have a clear picture of the end-to-end process, the roles and users involved in the process, a list of systems and tools that the process uses, and a good estimate of the volumes that go through the process (this could be hourly, daily, weekly, etc.). You also know how long each step in the process takes, along with any other performance metrics that may be relevant. You have also identified any known constraints and risks to automating the process. Also, because of your discussions with the SMEs and business partner, you know which steps in the process the business wants automated; this could be all or part of the process. And there may be other important information that you now know due to your discussions.

All this information is required to move to the next step, opportunity assessment, which will estimate RPA value and RPA feasibility as a solution.

You will note that you have been in contact with a number of different people throughout this phase of the process. A representative of the business made the initial request; you had at least one and possibly more interactions with them; then you contacted SMEs and possibly someone from your risk and compliance area. At this stage, you should contact your Center of Excellence (COE) if your organization has one. If not, who will be responsible for deploying the Bot? That person or department should be made aware of what is being proposed and a very general deployment date. Getting this involvement early on will serve you well as you proceed through the process of project initiation and development. You will see how in subsequent chapters.

Let's look at these concepts in a practical, although fictional, example, as shown in this case study.

Case Study: Client Address and Phone Number Update

For any client-facing organization, it is important to have the clients' most up-to-date contact information, to provide them with timely customer service (e.g., sending invoices, bills, or service change updates and marketing material such as promoting sales events, new services or products, etc.). In this hypothetical situation, a client submits an "address change" form via email to the hypothetical "MyRPA" company's customer service email inbox. There could be different types of emails arriving to that inbox; thus, customer service personnel review the emails and direct them to the appropriate team queue for processing.

In this case, the request went to a team A, which specializes in client profile management tasks. Team A consists of six customer service specialists who access the team's backlog folder every morning and process the requests in first-in, first-out order.

After a successful change, a customer service specialist (CSS) sends a confirmation email to the client.

After a process quality audit, the team's leadership discovered that a high number of address change requests were processed incorrectly due to human error, which causes a high number of client complaints along with high operational costs due to further investigation and the rework required to correct those errors. The leadership wants to explore an opportunity to automate that process to improve overall process quality and redirect their resources to other areas which require manual intervention.

****

For clarity, we need to distinguish between a process flow and a process map.

The following is a high-level *process flow* describing the process in the case study:

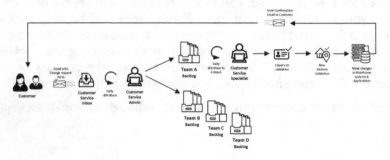

The following is a high-level *process map* for the case study (please see Appendix 2 for a template):

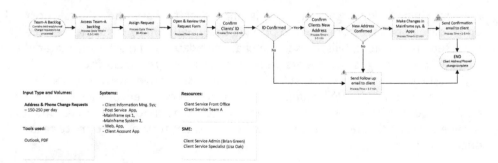

The request form for this example might look like this:

RPA Opportunity – Request Form

Background			
Initiative Name	Address and Phone Number Change Process		
Sponsor	Mary Smith		
Champion	John Doe		

How Does the Process Currently Work?
A client change request email containing new client information – address and/or phone number – is directed to team A for processing by a client service specialist (CSS). The team consists of five FTE working on those requests. It takes about 10–15 minutes per request for a CSS to process it when the address contains no errors and about 20 minutes when investigation or further follow-up with the client is required. The team leadership sees the opportunity to automate that process at least partially and redirect a few CSS to other areas where we are experiencing longer service time and higher overtime numbers. In addition, after auditing the process requests, we found that there are a high number of human errors in phone numbers, having the wrong client's address changed, errors in the new address, or client information not being changed on all the required systems and applications. This causes frustration among clients and increases the overall operational cost of customer services; since further investigation is required, we are not able to easily reach the client.

What Is the Goal?
We would like to improve the process quality and reduce the number of CSS working on those tasks. The ask is to automate the end-to-end address and/or phone number change process. Automation should be able to validate the clients' credentials, change the phone number, and/or validate and then update the information in the mainframe systems with the new address.

Common Business Drivers/Business Criticality
Operational cost, quality, improved customer experience, company image and reputation

Impacted Business Areas		Impacted Systems/Tools	
Client Service Teams		Outlook, PDF forms, Client Information Management System, Mainframe System	

Opportunity Type	Duration of the Solution?	Estimated Operational Benefits (FTE Hours)
Annual	5 years	7 FTEs, about 11,830 FTE hours

Priority		Desired Delivery Date	
Must do		10/01/20xx	

Form Completed Date	05/24/20xx	Prepared by:	Judith Wilson

During discovery phase, the process analyst engaged the SMEs (subject matter experts) and was able to capture the end-to-end (E2E) process and identify critical information, including

- Input types and volumes, tools and systems involved in the process, and resources executing the work

All that information was compiled in the process map, shown above.

In addition to the process map, a detailed description of the process steps was created (please see Appendix 3 for a template). The benefit of having this document is to provide any required details that might be critical to decision-makers and developers further down at next steps.

Client Address and Phone Number Update

1. Access Team A Backlog

Input	This is the start of the process.
Activities	Team A members access the backlog of requests.
Output	The backlog has been accessed.
Roles	Data update clerks on team A.
Notes	There is no separate "input" here. This is the trigger that starts that process and is done daily in the morning.
Metrics	Number of new requests per day.

2. Assign Request

Input	The backlog has been accessed.
Activities	Each request is forwarded to a clerk for processing.
Output	The requests have been forwarded.
Roles	Data update clerks on team A.
Notes	There must be some awareness of workloads.
Metrics	Number of requests assigned per day.

3. Open and Review the Request Form

Input	The requests have been forwarded.
Activities	The request form is opened and the request is understood (address change, phone number change, etc.).
Output	The work required for the request is known.
Roles	Data update clerks on team A.
Notes	
Metrics	

4. ID Confirmed?

Input	The work required for the request is known.
Activities	The account is researched to assure that the number on the request pertains to the name of the requestor.
Output	It is known if the information in the request pertains to the requestor (e.g., account number and name match).
Roles	Data update clerks on team A.
Notes	
Metrics	

5. Send Follow-up Email to Client

Input	Two possible inputs: The identification is not confirmed (a "No" decision from Step 4). The new address is not confirmed (a "No" decision from Step 6).
Activities	An email is sent to the client, requesting clarification.
Output	An email has been sent to the client.
Roles	Data update clerks on team A.
Notes	A template is used: one for unconfirmed ID and a different one for unconfirmed address.
Metrics	Number of emails that must be sent per day.

6. Confirm Client's New Address

Input	Client identification has been confirmed (a "Yes" decision from Step 4).
Activities	The necessary steps are taken to assure that the client's new address is accurate.
Output	It is known if the information in the request pertains to the requestor (e.g., account number and name match).
Roles	Data update clerks on team A.
Notes	
Metrics	Percent of unconfirmed addresses per day.

7. Make Changes in Mainframe System and Applications

Input	The new address has been confirmed.
Activities	All systems and applications that access the client information are updated with the new information.
Output	The systems and applications have been updated.
Roles	Data update clerks on team A.
Notes	
Metrics	

8. Send Confirmation Email to Client

Input	The systems and applications have been updated.
Activities	Using the template, an email is sent to the client, confirming the update of the information submitted.
Output	The confirmation update has been sent.
Roles	Data update clerks on team A.
Notes	
Metrics	Number of confirmation emails sent daily.

After this information is complete, you will schedule the Governance Committee meeting. If the project is approved at this point, you are ready to move to opportunity assessment. At this step, you will look in detail at the request. The following chapter describes this step.

Opportunity Assessment

Business leaders must know that RPA cannot automate every process. Hence, selecting appropriate processes to use RPA can be a complicated task. Opportunity assessment's goal is to help decision-makers choose processes that will be suitable for RPA. Selecting processes that don't fit the criteria will cause unnecessary expenses and create a drag for the team by consuming their resources on non-value-added activities. On the other hand, processes that lend themselves to automation will utilize available technology to its maximum potential and bring much higher return on investment.

■ **Tip** Keep in mind that RPA is a phased, iterative program. What may seem to be a good fit for RPA at the "opportunity identification" stage does not mean the request will be completed through to deployment.

The tasks within opportunity assessment begin once the initial request for automation has been received and understood. This is where the request is explored in some depth; a variety of information is obtained and examined.

Sometimes an idea is submitted with very little information; in those cases, the discovery phase described in Chapter 4 is performed before opportunity

© Robert Fantina, Andriy Storozhuk, Kamal Goyal 2022
R. Fantina et al., *Introducing Robotic Process Automation to Your Organization*,
https://doi.org/10.1007/978-1-4842-7416-3_5

assessment. Please note that whether or not the discovery phase is implemented depends on the amount of information received with the initial request.

Please note also that a request introduces an RPA opportunity; that opportunity may or may not become a project.

At the conclusion of the opportunity identification (which may or may not include the discovery phase), the governing body, described in Chapter 3, reviewed the documentation and then met to discuss whether or not to move forward at this time with the request. If a "Go" decision was made, the request is approved to move to the opportunity assessment phase.

The following is the tollgate decision template (see Appendix 7):

"Go/No-Go/Wait" Tollgate Decision

Purpose:

To provide all key stakeholders with an interactive opportunity to review the output of the opportunity assessment due diligence efforts and to render a decision on the merits of the RPA opportunity to proceed/not proceed or wait until specific conditions are met.

Decision	Reason	Next Steps

Approved by: Date:

RPA Dev. Lead _____

System Architect _____

RPA Analyst _____

Process/Business Owner _____

Details of the form:

> Decision: This is either "Go," "No Go," or "Wait." "Go" indicates that the governance body has determined that moving to the next phases is beneficial; "No/Go" indicates that, for any of a variety of reasons (e.g., insufficient ROI, complexity, etc.), the project will not continue. "Wait" indicates that there is some additional information required before a "Go" of "No Go" decision is made. That will be described in the next column.

Reason: This explains why the decision was made and, in the case of a "Wait" decision, what is lacking.

Next Steps: For a "Go" decision, this would indicate the next phase. For a "No Go" decision, this would indicate that the requestor is to be notified and who specifically will notify them.

■ **Tip**   Approval to proceed occurs several times throughout the life cycle. Approval for opportunity assessment does not mean approval for deployment.

Throughout this chapter, we introduce the required concepts and provide real-world examples to clarify them.

At this step, a high-level understanding of the request is obtained. After some preliminary information is discovered, analyzed, and documented, the team presents it to the Governance Committee with their recommendation. We detail what should be considered as benefits, which will include the potential return on investment (ROI), if the project is necessary to meet some competitor's initiative, if it will satisfy a new government regulation, if it is required to enhance customer experience, or any of several other benefits.

As noted, there are cases where, even from this initial view, it can be determined that the request is not a good candidate for RPA, for any of a variety of reasons.

At the conclusion of the phase, the Governance Committee will meet to determine if the request should proceed to the next phase (solution design – see Chapter 6).

This decision will be based on many factors, which include time/cost savings, competitive need, regulatory requirements, improving quality, enhanced customer services, and other factors depending on the request and the needs of your organization.

In this chapter, we discuss the following artifacts that are produced in this phase (see the Appendix for templates):

- The opportunity overview, containing a summary of the current situation, objective, desired state addressing what process steps are proposed to be automated, and benefits estimations. At this point, only a high-level estimate of costs and benefits is obtained.

- An end-to-end, current-state process map, as we discussed in Chapter 4, picturing actual process flow and the level of detail required at this point (e.g., process volumes, systems/applications involved, roles, etc.). If,

when it is time for the "go/no-go" decision, the project is
approved to move forward, a more detailed, and a future-
state, process map will be required. But that will be
discussed in a later section. We also include a template
for the process map.

- A more detailed explanation of each step within the
  process. A template for that is also provided. This is not
  always necessary, and we will discuss how to determine if
  it is.

- A feasibility assessment.

- A risk assessment.

Opportunity Overview

A major component of the opportunity assessment is the opportunity
overview. This is an at-a-glance view of the request, which includes such
information as an overview of the need, desired solution state, and the
estimated benefits of automation (e.g., ROI, benefits could be qualitative and/
or quantitative), which can be financial, or might include maintaining a
competitive edge, satisfying legal requirements, or several other factors. Much
of the information contained therein is extracted from the request form and
can be obtained during discovery. Please see Appendix 6 for a template.

In other words, the opportunity overview is an executive summary that
provides the Governance Committee members with a thorough, but high-
level, overview of the opportunity, the possible solution, and its complexity,
risks, and benefits.

■ **Tip** Always remember that the templates included can be tailored to meet the needs of your
organization.

These are components and details of a typical opportunity overview:

- Name of project: Self-explanatory. Remember, this
  request is not yet a project, but this is the title that will
  be given to the work which may result in a project of this
  name.

- Prepared by: This is generally the business process analyst and any subject matter experts (SME(s)) who might
  be involved, along with their titles and departments.

- Opportunity description: Information from the "current process description" on the request form is the basis for this. In the opportunity overview, the information is refined from the basic notes on the request form to make a more coherent description of the current situation and the problem as it now exists.

- Opportunity statement: This should be short; a sentence should suffice. Here, a concise summary of the problem is stated. For example, "The current process manually performed results in unacceptable delays and repeated, time-consuming rework due to errors."

- Objective: Write what the requestor wants accomplished in simple language; this should be a general description of what the Bot should do. Express in your own words what you want it to do. For example, "Bot will scan orders and produce invoices with customer address included..." or "Bot will read customer email and extract information to make required updates in the systems."

- Assumptions and constraints: In this section, briefly include anything that is believed to be true and is pertinent to the project. For example, "Assumption: System 123 and System ABC are able to communicate." For constraints, list anything you think may prevent the easy build of the Bot. For example "Currently, there is no way for accounts ending in L to be read by system XYZ." Further explanation is not required here.

- Current state: The process map, at a sufficient level to show the steps to be automated, is placed here. Please see Appendix 2 for a template.

- Operational benefits estimation: Some preliminary benefits were listed on the request form, but more specific information is detailed in this phase. It should clearly, but without great detail, show the advantage(s) of the requested automation. Is the savings (ROI) sufficient to justify the cost of automation? Are there important competitive or regulatory issues that the requested automation will address? These considerations must be included.

- Proposed solution/process change: Note here exactly what is to be automated. For example, "Automate Steps 3–7 on the previously referenced process map." It could also be descriptive; for example, "Automate the identification of accounts over 90 days past due, the sending of notices, and notification to the account rep. that the past-due letter has been sent."

■ **Pause and Consider** An opportunity overview is a high-level assessment of the request to automate a process. Management, who will review the opportunity overview and make a decision about the request, does not need or want the same level of information that the analysts, developers, testers, etc., will need. Adding extra information may seem like a good way to "cover all bases" but restricting the opportunity overview to just the minimum facts required to make a decision will serve much better.

Current-State Process Map

Automating a process means using the existing process steps and simply changing them from manual to automatic performance (note: in some cases, it may be necessary to optimize the process prior to automation). In order for the Governance Committee to make a decision on automating a process, and for developers to create the solution, the process must be understood. The current-state process map is one of the main tools by which this is done.

At this stage, the map should be fairly high level; it isn't necessary to state, for example:

> "Access screen X; find the 'account number' drop-down; hit the drop-down button; select 'individual insurance'; click ok; enter customer last name; click ok; enter customer first name; etc."

> The step could simply say "find customer account number." The details of the steps will be defined, probably by video recording someone actually performing the steps in the process, at a later stage in the project. What is required now is a solid understanding of the end-to-end process at a high level.

Feasibility Assessment

Another major factor of opportunity assessment is feasibility. At this point, "feasibility" only relates to the high-level view we currently have. Does it appear, based on the limited information thus far available, that the proposed process is suitable for automation? In other words, is RPA a reasonable solution to the presented opportunity? The goal of the feasibility assessment is to identify, as early as possible, requests that are not suitable for RPA and avoid spending unnecessary resource hours. The feasibility assessment, with a recommendation by the RPA team, will be a major part of the Governance Committee's decision-making process about the requested automation.

Remember that if the project is approved to move to the next stage, more information will be obtained during that stage, and there will be another Governance Committee checkpoint, at which the new information will be evaluated. It is possible that a project that was seen as "feasible" when very little information was available may be viewed differently with additional information and could be cancelled. It is very important to reflect and learn from the information or details that lead to cancellation of the RPA development and incorporate that into the feasibility assessment tool to minimize the RPA cancellation rate. By achieving a low cancellation rate, you would decrease operational cost (higher-technology ROI), increase delivery velocity, and improve business operations' trust and user satisfaction. Additionally, with the evolving capability of digital automation, some previously impossible solutions might become feasible; therefore, you need to make appropriate adjustments to the feasibility assessment tool over time, as you gain more experience with RPA.

■ **Pause and Consider** Why is cancellation at a later stage not a "failure"? Remember, the Governance Committee exists to evaluate the current information. At "opportunity identification" or "opportunity assessment," only basic information is known, and the Governance Committee isn't giving approval for the entire project, just for the next phase, wherein additional information will be discovered. The Governance Committee will then evaluate that information.

At this stage, there are six basic questions to ask.

Feasibility Assessment

<Name of Project>

	Questions	Answers
1	Is the process well defined?	
2	Is the process stable (very few "exceptions")?	
3	Can exceptions be handled manually?	
4	Are inputs in digital format?	
5	Can required data be input without human intervention?	
6	Are potential changes to roles and processes acceptable to management?	

Approvals:

<Name> <Role> _____ Date: _____

<Name> <Role> _____ Date: _____

<Name> <Role> _____ Date: _____

<Name> <Role> _____ Date: _____

(See Appendix 4 for a template.)

1. **Is the process well defined?** Here, you need to determine how repetitive the process is. Basically, are the exact same steps performed each time the process is run? Are the steps well-known? Are there a limited number of exceptions? This is one characteristic that makes a process a good candidate for automation.

2. **Is the process stable (very few "exceptions")?** When the process is invoked, are the steps that are taken based on clearly established rules? Are the steps the same within the process?

3. **Can exceptions be handled manually?** For any process, a certain, limited number of transactions might need some additional steps. Perhaps an account number was entered incorrectly, and the account must be looked up by last name. A customer might make a special request within a more common request. Can these transactions, if within an automated process, be sent to a mailbox for manual handling? Is there a process for handling these exceptions now?

4. **Are inputs in digital format?** Do the inputs currently come into the process via an electronic mailbox? Or are they calls from clients or others that must be input by the call receiver? If so, can the input created by the call receiver be input into a system that the Bot can access?

   Also, are all inputs Excel files? Word Documents? PDFs? Etc. How easily might they be read by a Bot?

5. **Can required data be input without human intervention?** A Bot does not think; it looks for exactly what it is told to look for and handles what it finds accordingly. If inputs require human judgment, beyond a simple decision of if it can be handled by the Bot, that process is not a good candidate for automation. Tasks that require little to no judgment and have low exception rates are good candidates for RPA.

   For example, an electronic mailbox may receive 300 emails per day, and 100 of them can follow a standard process. It may take a person to manually review the 300 emails and forward the 100 that can follow a standard process to the Bot. But once there, the Bot can take over.

   If, however, the mailbox receives 300 emails per day, and each requires that a person read them and look up a variety of different information that may be available from a variety of different and ever-changing sources, then this process would not be a good candidate for automation.

6. **Are potential changes to roles and processes acceptable to management?** One advantage of automating processes is that staff will be freed up for other responsibilities, changing their roles. Also, when the process is being investigated for automation, efficiencies may be determined that will change the process before it is automated. These and related changes must be accepted by the process owner, in order to move forward.

■ **Pause and Consider**   Are there other considerations in your organization that should be added to a "feasibility assessment" template? Remember, these basic questions are fairly standard, but there is no "one-size-fits-all." You may start with these and use them for the first few RPA projects, but your "feasibility assessment" template, along with all the others included herein, may evolve over time to meet the specific needs of your organization.

Risk Assessment

For any new RPA opportunity, a preliminary risk assessment must be performed. Most project leaders are well versed in risk assessment activities, but there are some risks that are peculiar to RPA, and we detail them in this section. It's important to know that, as with any project, in this phase (opportunity assessment), only preliminary risks will be identified, and more will be discovered when and if the project moves forward. An early involvement of your organization's risk and compliance group (it may have a different name in your company) is beneficial. That group can assist in identifying risks early on. If they are identified later, it might necessitate costly rework or even cause the project to be cancelled. This would result in an unnecessary waste of time and money.

Overall, the purpose of the risk assessment is to list what might go wrong during the creation of the Bot or during implementation. For example, is there a likelihood that certain applications that must be accessed by the Bot will be difficult to access? If so, how much will that increase the time to develop the solution? Another example: How great is the possibility that the Bot will misread information? These kinds of risks must be identified as early as possible to help the Governance Committee make informed decisions.

Writing in Forbes on June 18, 2019, Naveen Joshi noted the following four (4) major risks:

1. Incorrect process selection
2. Technical issues
3. Lack of communication
4. Security[1]

The risk assessment document will be used by designers and developers to create the automation that will mitigate those risks. While a thorough risk assessment is required, it must be remembered that at this point, there is still limited information known about the project and almost nothing about the solution. We only have a clear understanding of what the objective is and what the "ask" is (what does the requestor actually want). In other words, we know what and where, but not how. Additional risks will be identified as the solution development progresses.

Start with this form (there is a template in Appendix 5). An explanation of each category follows the template.

---

[1] www.forbes.com/sites/cognitiveworld/2019/06/28/leverage-rpa-but-plan-for-its-inherent-risks-too/?sh=191b34d811d1. Accessed on April 9, 2021.

Risk Assessment Checklist

<Name of Project>

Risks	Type	Is There a Mitigation Plan	Mitigation Plan and Responsible Party

Prepared by:

<Name> <Role> _____ Date: _____

<Name> <Role> _____ Date: _____

Approvals:

<Name> <Role> _____ Date: _____

<Name> <Role> _____ Date: _____

(See Appendix 5 for a template.)

1. Risks: In this area, briefly and succinctly identify the risks. One that is standard for RPA initiatives is the following: "The Bot fails to function."

2. Type: Generally, the "risk type" is one of the following categories.

    i. Compliance

    ii. Error

    iii. Financial

    iv. Operational

    v. Reputational

    vi. Resource

    vii. System

    viii. Technology

    ix. Security

Complete this area to the best of your current knowledge. Remember, this is a very early stage in the project, and additional risks will be identified as more information about the business need, the technology, etc., is obtained.

■ **Tip**   Solicit input on risks from a variety of subject matter experts to assure that you assess and identify potential risks RPA could expose or create. Involve your "risk and compliance" group in this.

---

3. Is there a mitigation plan? This could be "Yes," "No," or "N/A." "Yes" means that there is a mitigation plan for this particular risk. For the standard risk mentioned before, this will be "Yes."

   "No" means that there is no mitigation plan for this particular risk, but one must be created.

   "N/A" means that there is no mitigation plan, but the decision is made to simply accept the risk. This may include such things as accepting maintenance costs, creation of the technical debt, or accepting the fact that X% exceptions will require manual handling.

4. Mitigation plan and responsible party: In this box, you will succinctly express what must be done about the risk. If there is a mitigation plan ("Yes" from the column titled "Is There a Mitigation Plan"), briefly describe it. For example, for the standard risk mentioned previously, the "Mitigation Plan" is usually this: "Revert to manual handling until the Bot is repaired."

If there is no plan ("No" from the column titled "Is There a Mitigation Plan"), briefly describe the steps required to create one. This might include any of the following (among others): "Confer with corporate risk management"; "Obtain input from SMEs"; "Research industry journals."

Also in this section, put the role (not the name) of the person responsible for handling the risk. For the standard risk, this is the RPA team lead, along with the role of the person responsible for the process.

At the bottom of the page, include your name and the names of anyone who worked on the risk assessment, the date of the completion of the document, then the names and roles of the approvers, and the date of approval (approvals are generally provided via email, to maintain a "paper" trail).

▓ Pause and Consider At the point of RPA transition to production, all identified risks must have a mitigation plan with an assigned owner. The plan may be simply accepting the risk, but that must be documented, and that acceptance must have an owner. How will you identify risk owners and document their acceptance of the risk and/or mitigation?

Remember that the purpose of the opportunity assessment is to evaluate the request or RPA candidate for suitability and value prior to investing resources for solution development. Human nature is to jump to execution mode, and it may seem that it is a faster and more productive way, but it is more effective and more efficient to have proper assessment or evaluation of RPA candidates, especially at the early stage of RPA introduction to the organization. The amount of time spent up front will decrease as RPA culture and the technology architecture environment mature. But the introduction of RPA to an organization is not the time to cut corners.

Once all documents required for the opportunity assessment phase are complete, you are ready to proceed to the next Governance Committee meeting.

▓ Tip If a particular artifact does not bring value to your organization, don't use it. But be careful before eliminating a step in the early stages of your RPA initiative. Let experience guide you.

Governance Committee Meeting at the Conclusion of Opportunity Assessment

Now that you have looked more closely at the request by completing the documents shown earlier, it is time for the Governance Committee to determine if it is still feasible to proceed to the next step. The RPA team's recommendation will be important in the Governance Committee's decision-making. Making the decision to either move forward to the next phase of the project (solution design; please see Chapter 6), postpone the project, or cancel it altogether cannot be made lightly. There are a wide variety of impacts to any of those choices.

As mentioned earlier, there are several times along the project life cycle that the Governance Committee will evaluate the findings obtained from a phase (e.g., opportunity assessment, solution design) and determine feasibility. The RPA team will combine the previously referenced documents into a single document, schedule the meeting, and include the link to the document (if not everyone on the Governance Committee has access to where the document is stored, simply attach it to the email).

Prior to the Governance Committee meeting, the Governance Committee members should all have read the document and should come prepared to ask questions. Realistically, this is often not the case; the Governance Committee members may not have reviewed the document prior to the meeting. The process analyst/designer or another RPA team member must be prepared to give a very brief, high-level overview of the request. This overview will include a high-level description of the current process. The person presenting will also discuss the problem that the automation is meant to resolve and the benefits of automating the process; that is, will it achieve the requested benefits? Also, they will discuss the feasibility of implementation: does it seem as if the manual process is a good candidate to be automated in terms of cost, complexity, etc.? Lastly, they will present the risks both of automating the process and any risks associated with not automating the process. This might include the potential to lose a competitive advantage.

All the information from this brief summary can be obtained from the prepared documents, with most of it coming from the opportunity overview.

Based on either Governance Committee members' review or the brief summary provided, there are three possible outcomes:

1. The project is approved to move to the next phase (solution design).

2. The project is put on hold, pending some additional information. Depending on the additional information required, a second Governance Committee meeting may or may not need to be held. If the information is minor and advising the Governance Committee members via email is sufficient, a second meeting need not be held. If email communication is all that is needed, each member of the Governance Committee must signify their decision (approve, obtain still more information, or "pause") by return email.

You will see that, as the culture matures, there will be fewer times when additional information is required. As the RPA team learns what information the Governance Committee requires, that information will be obtained prior to presenting the opportunity to the Committee.

In some cases, there will be requests that are not currently feasible, but could be made feasible. For example, it may be that all inputs are manual, and that is the only thing preventing the process from being a good candidate for

RPA. But if the process owner is able to make them electronic, then the process could be automated. Following the Governance Committee meeting, the process stakeholders will be advised of any such situations, so they can determine if they want to make the changes required to qualify for automation.

3. The project is "paused." This could be because while automation would benefit the department, the cost of automation would far exceed continuing to do it manually. It could also be due to extreme complexity or any of a variety of other reasons.

There may come a time when, due to a light workload, changing priorities, or other situations, the project will move forward. Each individual RPA team must determine if all "paused" projects will be reviewed annually, semi-annually, etc.

4. The project is cancelled. There are times when a project is simply not a good fit for RPA. The business stakeholders may certainly have an issue with a manual process, but RPA isn't the solution. In these situations, whenever possible, refer the business stakeholders to some other possible solution. Most organizations have a variety of applications, and often these can provide some improvement to the identified issue.

■ **Pause and Consider**   Requests are "paused" when they will be reviewed again in the future. It is not good to cancel a project that the requestor and their department have been excited about. How can you best advise the requestor that their request has been "paused" or cancelled? Remember, you don't want to discourage a requestor from submitting another process for automation.

Now let's look at how the concepts in this chapter would apply to the fictional case study we introduced in Chapter 4.

Case Study Example

We would start with an opportunity brief. This document should not exceed two pages (please see Appendix 6 for a template). For our example, the opportunity brief might look like this.

Address and Phone Number Change Process

Prepared by: David Johnson, Business Process Analyst, Operations Support, Business Process Automation

Current Situation:

Clients notify us by emailing their new address and/or phone number. These emails are reviewed by a client service specialist (CSS). The team consists of five FTE working on those requests. It takes about 10–15 minutes to complete each request, if address contains no errors, and about 20 minutes when investigation or further follow-ups with the client are required. Management sees the opportunity to automate that process at least partially and redirect some CSS to other areas where we are experiencing longer service times and higher overtime numbers. In addition, auditing has shown that there is a high number of human errors either in phone numbers, or wrong client's address was changed, or there are errors in the address.

Objective:

Automate the end-to-end address and/or phone number change process. Automation should validate the clients' credentials and change the required information in the mainframe systems and relevant applications.

Assumptions:

Automation can:

- Extract information from the request form.
- Confirm client's ID.
- Check/validate new address.
- Access mainframe systems and relevant applications.
- Store execution data for monitoring, analysis and reporting.

Constraints:

- Automation recovery and accumulated backlog process is less than 48 hours.
- CSS availability in the event that manual process is required.
- Address verification could be challenging.

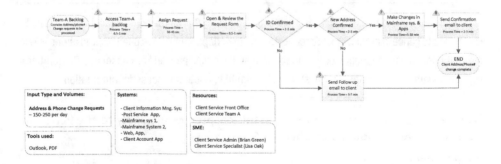

Input Type and Volumes:	Systems:	Resources:
Address & Phone Change Requests ~ 150-250 per day	- Client Information Mng. Sys; -Post Service App, -Mainframe sys 1, -Mainframe System 2, - Web, App, - Client Account App	Client Service Front Office Client Service Team A
Tools used: Outlook, PDF		**SME:** Client Service Admin (Brian Green) Client Service Specialist (Lisa Oak)

Benefits Estimation:

Processing requests: FTE hours = 9,375/year (5 FTE), which can be deployed elsewhere.

Eliminate rework.

Eliminate the cost of returned mail.

Proposed Solution:

The Bot will access the teams' email, open the form, confirm the client's ID, and validate the information it contains. It will then access the mainframe systems and relevant applications, make the required updates, and send a confirmation email to the client.

In case of information that cannot be verified, the Bot will redirect the request for manual processing.

(See Appendix 6 for a template.)

Next, we look at feasibility assessment. Are the basic conditions listed on the document (and discussed earlier) satisfied? Remember, it isn't necessary that all of them be satisfied immediately, but by the end of this phase (opportunity assessment), the team must be comfortable in knowing that they can be satisfied. For example, if the inputs to the process are not digitized, it must be confirmed by the process owner and SMEs that they can be.

Feasibility Assessment

Address and Name Change Process

	Questions	Answers
1	Is the process well defined?	Yes
2	Is the process stable (very few "exceptions")?	Yes
3	Can exceptions be handled manually?	Yes
4	Are inputs in digital format?	Yes
5	Can required data be input without human intervention?	Yes
6	Are potential changes to roles and processes acceptable to management?	Yes

Prepared by:

<Name> <Role> _John Williams, Tech Lead._____ Date: August 4, 202x

<Name> <Role> Mary Davis, Business Manager_____ Date: August 4, 202x

Approvals:

<Name> <Role> _____ Date: _____

<Name> <Role> _____ Date: _____

<Name> <Role> _____ Date: _____

<Name> <Role> _____ Date: _____

Next, we look at risks that the solution could potentially expose the operations to. We have discussed this topic previously, but the risks identified will be unique to each request. Remember to rely on SMEs and to include your "risk and compliance" department during this phase.

Risk Assessment Checklist

Address and Phone Number Change

Risks	Type	Is There a Mitigation Plan	Mitigation Plan and Responsible Party
Automation fails to execute.	Operational	Yes	Revert to manual processing until repaired; James Morris, CSS Assistant Manager and RPA team.
Wrong customer information updated.	Reputational	Yes	Establish client ID validation controls; Maria Mulchuk, Compliance Manager.
Client information cannot be confirmed.	Technology	Yes	Handle manually; James Morris, CSS Assistant Manager
Client information exposed.	Compliance	No	

Prepared by:

<Name> <Role> _John Williams, Tech Lead._____ Date: August 4, 202x

<Name> <Role> Mary Davis, Business Manager_____ Date: August 4, 202x

Approvals:

<Name> <Role> _____ Date: _____

<Name> <Role> _____ Date: _____

<Name> <Role> _____ Date: _____

<Name> <Role> _____ Date: _____

(See Appendix 5 for a template.)

Note that this document is still incomplete. There is still a risk for which mitigation plans have not been established. At this phase, that is acceptable, but those mitigation plans must be established if the request is approved to move forward.

After these documents are created, a Governance Committee meeting is scheduled for review and to make a decision either to proceed with development or not. Again, the recommendation of the RPA team will be a key input into this decision.

Let's now review the Governance Committee decision form.

Governance Committee Decision Form

Address and Phone Number Change

Decision	Reason for the Decision	Next Steps
Go	Automation solution is technically feasible. Process is repeatable and straightforward. Benefits meet ROI criteria. No concerns from system architecture.	Proceed with solution design.

Approvals:

<Name> <Role> Siva Khrishnak, RPA Dev. Lead _____ Date: August 4, 202x

<Name> <Role> Daniel Cook, System Architect _____ Date: August 4, 202x

<Name> <Role> Ryan Walthy, Ops Risk and Control _____ Date: August 4, 202x

<Name> <Role> Andy Hamilton, Business Owner _____ Date: August 4, 202x

(See Appendix 7 for a template.)

This form is, of course, blank entering the meeting. At the meeting, any questions the Governance Committee team members have are answered. In this hypothetical request, the decision was made to move forward (see "Go" in the "Decision" column), the rationale for the decision is listed in the "Reason" column, and the next phase – solution design – is listed in the next step column.

Conversely, the "Decision" could have been "Pause," with the reasons listed in the "Reason" column, and "Next Steps" would have provided additional explanation. It might have said "Once process owner confirms digitization of 90% of input, this will be re-reviewed," or some other description of what the "Next Step" would be.

Finally, the "Decision" could have been "No Go." The rationale would have been described in the "Reason" column, and the "Next Steps" might have been "Advise the customer that the ROI is insufficient to justify the cost of development at this time. Recommend that the customer speak with (name, role) about using (application name) to provide some improvement." That is, of course, if some alternative that would be helpful is known.

Once the project has been approved to move forward by the Governance Committee, you are ready to move to the "solution design" phase.

Solution Design

During conceptual design, requirements (functional and nonfunctional) are gathered. Templates are provided in the Appendix, and clear explanations are detailed in this chapter.

Conceptual design can be seen as the solution blueprint. A high-level determination will be made of the automation solution to include what the Bot will do, the systems it must access, the schedule on which it must run, etc. This will be further detailed, but in this phase, the basics will be determined.

Also during this phase, a solution design document is created. This document includes process and outcome, overall structure, logical components and interrelationships of both the current and possible future states. We provide a template, with a clear explanation (please see Appendix 12).

RPA can be introduced with either a conventional "Waterfall" approach or with a more Agile methodology. In the "Waterfall" approach, requirements and conceptual solution design are created and documented before development begins. More and more companies are adopting an Agile or concurrent engineering approach, where requirements and solution development activities are ongoing in parallel. Each of these approaches has pros and cons, and some organizations adapt them to a hybrid approach.

Regardless of the methodology selected, the tasks described within this chapter must be completed.

© Robert Fantina, Andriy Storozhuk, Kamal Goyal 2022
R. Fantina et al., *Introducing Robotic Process Automation to Your Organization*,
https://doi.org/10.1007/978-1-4842-7416-3_6

Once conceptual design is finished, a readiness checklist must be completed. This assures that all the components required for development (the next stage) are complete. A template with a detailed explanation is included.

Once the readiness checklist is completed, it is forwarded to the governing body for a decision on whether or not to move to the next phase, development. The governing body generally consists of the following roles:

- RPA technical delivery lead
- Architect
- Developers
- Others as deemed necessary

Pause and Consider   Names of roles differ from company to company. What are the coinciding roles within your organization? Also, consider what "others as deemed necessary" might be in your company. This could be a director, area manager, system analyst, or others. The governing body should be small, but should have the right people to decide whether or not to move to the next phase.

Please note that one of the advantages of this model is that work is done gradually, and there are very specific checkpoints for the governing body (decision-making body) to review the work and determine the continued feasibility of any project. A request that, during opportunity assessment, when only preliminary information is gathered, may seem promising in terms of ROI, competitive advantage, or any other factor may seem less so upon the deeper investigation done during conceptual design. This is true as the project moves from any phase to the next one. At any point, the governing body may decide that earlier evaluations have not proven true upon closer scrutiny and cancel the project. A project once given the go ahead in one phase, and later cancelled, does not indicate a failure in the earlier phase. This is the nature of RPA project progress.

Pause and Consider   In many companies, projects often continue even when stakeholders know they are no longer needed. Cancelling a project is seen as a waste of the money already spent and the failure of various decision-makers. How will you convince management that using this model of periodic evaluation of a project is beneficial?

In RPA, the requirements document is generally referred to as the Process Design Document (PDD). This is described in the section, 'Governing Body Approval', below. If your organization is good with requirements gathering and documentation, continue to use that process; details of a typical requirements document follow.

But if the process of gathering requirements in your organization does not exist or is not efficient, and you need to capture requirements for RPA only, then PDD is the preferred way of gathering and documenting requirements. See the section, Governing Body Approval, of this chapter for that information.

Requirements Gathering

We will start with a discussion about functional requirements and the template we recommend.

Please refer to the template in Appendix 9, remembering that it is simply a model. If it works for your organization, fine, but feel free to adapt it to your organization's needs.

1 Level 1 Details

    1.1 System

    What systems must be accessed, during input, throughput, or output? List them here.

    1.2 Goal

    In this section, briefly but clearly describe the goal of the project. This might include a short description of the current situation and why it is a problem, along with a brief description of what the outcome of the project will be and its advantages. This is NOT where the goal is defined; that was done in earlier phases. This simply describes it in more detail. It is important not to fall into the trap of redefining the goal here and moving away from the initial request. This description must satisfy what the process owner requested. They must approve this document, so the description must be accurate.

    1.3 Triggering Event

    In this section, list what event causes the process to start. For example, client seeks to change beneficiary information.

    1.4 Triggers

    This is the specific action that causes the process to start. To follow the example in Section 1.3, the event is the client wanting to change beneficiary information. The specific action, or trigger, might be the receipt of an email from the client advising of the new beneficiary information. Or it might be notification from an advisor to make the change on behalf of the client. All possible triggers should be listed.

1.5 Precondition

In order for the process to start, what circumstances must exist for the Bot to begin? For example, the Bot may need to have access to certain software, the request may have to be in a particular format, etc. List all required preconditions here.

1.6 Post-condition

List here everything that is changed or produced as a result of the process. In the example earlier, post-conditions might include the following:

- Beneficiary information has been changed.

- Client has been notified that the change has been made.

- Advisor has been notified of the change.

- A report has been issued, listing all changes within the last day.

Tip Talk to subject matter experts (SMEs) to fully understand triggering events, triggers, and pre- and post-conditions.

---

2. Level 2 Details

2.1 Input Parameters

Parameters Name	Description/Value	Mandatory?	Meaning if omitted (only enter if mandatory = no)
What is the input to the process? It could be an email, the completion of another process, or any number of other things.	Briefly describe the input.	Is it mandatory or not?	
Insert as many input parameters as there are. Add extra rows as necessary.			

2.2 Output Parameters

Parameters Name	Description/ Value	Mandatory?	Meaning if omitted (only enter if mandatory = no)
What does the process produce? List its name here. Include all outputs. Add extra rows as necessary.	Briefly describe the output.	Is it mandatory or not?	

3. Level 3 Details

3.1 Main Flow Steps

In this section, list the steps of the normal process. For example, if in 70% of the cases, there is one way the process flows, list those steps here. In using the same example as before, the steps might be as follows:

1. Email received

2. System ABC accessed

3. Account number found

4. Beneficiary changed per instructions

5. Email sent to client

6. Email sent to advisor

7. Information added to daily report

3.2 Alternative Flows

There are some inputs that will have slightly different steps, but can still be handled by the Bot. List those process flows here.

3.3 Exception Flows

There may be some circumstance that the Bot will be unable to process. Using the example earlier, this could be because the account number couldn't be identified, or the current beneficiary name was wrong, etc. Explain what happens in those circumstances. Using the same example, the flow might be as follows:

1. Email received

2. System ABC accessed

3. Account number not found

4. Email sent to processing clerk for manual handling

5. Adding information to an exception report

■ **Pause and Consider** In order for the developers to create efficient Bots, this information is vital. How will you get it? Who do you need to speak with to obtain it and then validate it? Remember, spending the time now to get it right will prevent costly rework later.

We will now discuss **nonfunctional requirements** (see Appendix 10 for a template).

"Nonfunctional Requirements (NFRs) define system attributes such as security, reliability, performance, maintainability, scalability, and usability. They serve as constraints or restrictions on the design of the system across the different backlogs."[1]

1. Automation service standards: This is a "boilerplate" section that simply states the date that current standards were set and by whom (example: COE). If there are no established standards at your organization, please disregard this section.

2. Privacy, data retention, and purge requirements: Include information on protecting customer privacy and also where the data will be stored. Also indicate when the document can be deleted.

Process Flow

Include a process flow for visual clarity. Using the same example, the process flow might resemble this:

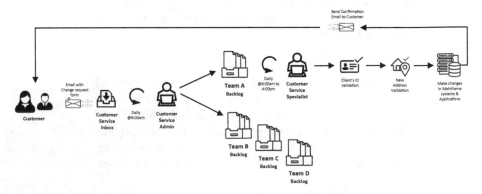

---

[1] www.scaledagileframework.com/nonfunctional-requirements/.

The process flow is an at-a-glance view of the process from end to end. It enables management to have the necessary understanding of the process and assists designers in knowing what aspect of the process can be automated and how best to design the automation.

■ **Tip** The terms "process flow" and "process map" are often used interchangeably. Please keep in mind how they are used in this book.

Screen Captures

In order for the developers to create the Bot, they must know exactly how the process functions. Remember, RPA is to automate an existing process. While there might be some opportunities for improving the process itself during automation, no major process changes will be made. If they are required, the process is not a good candidate for RPA.

■ **Pause and Consider** During this phase, the request will be examined in more detail. Is it still a good candidate for RPA? Remember, at the end of this phase, there will be another governing body meeting, and you will be asked for a "go/no-go" recommendation. You may have recommended "go" at the conclusion of opportunity assessment, but following a deeper look, you may recommend "no go." That is perfectly legitimate.

In order to get screen captures, the most effective method is to record a person actually performing the process manually. This can be done through any of a variety of online meeting apps or, lacking anything else, using a cell phone.

After identifying the correct SME, schedule a meeting with them, and record the entire process. After the meeting, grab screenshots and place them in a word document, using arrows to indicate selections, and have a brief description of what is happening on each screen. Remember, in order to correctly automate the process, the developers must see every last step in the process.

Finally, include a screenshot of a typical output. What is it that the process produces? Is it a report in Excel format? A letter to a client or advisor? Whatever it is, include a screenshot of it.

■ **Pause and Consider**   Your company may have standards that must be followed for development. Your new RPA program must conform to those standards. If there is a Center of Excellence (COE) in your company, those standards may be housed there. Bring in a COE representative early in the process, to assure that you have all the necessary steps in place to satisfy company requirements.

Planning

Formal planning begins once requirements are gathered. For a "Waterfall" methodology, this means that they have been fully captured, documented, and validated for the entire project, start to finish. If you are using an Agile methodology, or a hybrid of both, you must determine at what point you have requirements sufficient to begin planning.

Once you have reached the point where requirements have been sufficiently captured to satisfy the methodology used (Waterfall, Agile, or a hybrid), planning is the next step. Planning is a vital step in the process of automation. The old adage "If you fail to plan, plan to fail" is certainly true with RPA.

In planning, the first step is to schedule meetings with the pertinent stakeholders. Most of these will have been identified by now as you have talked to the business owner(s) and subject matter experts. Assure that you have all the necessary stakeholders identified. This can be validated in a brief discussion with the business owner(s). This initial planning meeting should include product (business) owner(s), key SMEs, a business analyst, developers, and/or a deployment team member.

The main purpose of this first formal planning meeting is to assess the size and complexity of the process for which automation has been requested. With the key stakeholders who will attend this meeting, you will be able to determine if the level of requirements detail collected is sufficient to start development or if additional detail is needed. This is a key outcome of this meeting.

Some information may be reviewed at this meeting, but if additional requirements are needed, they should be obtained first. You want to be sure to discuss what applications the automation is going to work with and understand if these applications are used in any existing automations. If they are, identify which automations are using the applications. You also need to determine what access is required to use these applications. You may need to engage with your organization's security team to get the right authorization. Lastly, you should determine if a proof of concept (POC) is required in order to determine technical feasibility.

If additional requirements are needed, or these questions cannot be answered at this meeting, do not proceed. Obtain the missing information, and meet again to finalize these issues.

For a Waterfall methodology, once the requirements are completed, reviewed, and signed off by the required stakeholders, the development team will meet regularly.

If using an Agile methodology, once sufficient information is obtained and the user stories are developed, the Agile project management process begins.

Pause and Consider The development team will meet "regularly." What will "regularly" mean in your organization? For an Agile team, what meetings additional to the daily stand-up will be required? If using a "Waterfall" methodology, how often must you meet? It is likely that your team may require more meetings at the start of the project and fewer as the project progresses.

During initial development team meetings, requirements must be divided into smaller tasks, functions, or stories. During these meetings, the team will identify dependencies and decide the sequence of development.

Once this information is developed, and the team has a good idea of complexity and a timeline, it is time to once again review the project with the governing body (governing body). The team will briefly present the information they have identified and make a recommendation on whether or not to proceed with the project. Remember, as more information is gained, a project may be validated, or it may be found to not have the promise that it initially appeared to have.

Tip Remember, a project cancelled at this point does not indicate failure! Based on your earlier knowledge and recommendations, the governing body has agreed to proceed. If new information has come to light that causes you to think the project is no longer feasible, that just demonstrates that the RPA process itself is working as it should.

Governing Body Approval

Tip We refer to the governing body as the "governing body," to indicate that it is the "gatekeeper" to the next phase. Use whatever terminology fits your organization.

As with all the phases, if the governing body decides not to proceed with the project to the next phase, the business owner will be advised of this decision and the reasons for it. If, during discussions with the team, any other possible solution was identified (e.g., a new software tool, a significant improvement to the manual process, etc.), the business owner should be informed of those possibilities.

If the governing body has approved moving to the next phase, the development team now starts working on developing process design documents (PDD).

As mentioned earlier in this chapter, PDD is the preferred method of documenting requirements for RPA projects. If your organization doesn't have an effective method of gathering and documenting requirements, use the following method for your RPA projects.

A process definition document is fairly straightforward. It answers some basic questions to assist designers and developers in automating the process it describes. It conveys what the process does, its inputs and outputs, the steps currently performed to accomplish the process, and any decision points in the process. Regarding any decision points, you need to determine if they can be automated. If not, they will need to be sent for manual handling.

The PDD also conveys who owns the process and who actually performs it. The document will also specify how often the process is performed: it might be hourly, weekly, monthly, on demand, or any other configuration. Also, what systems or applications interact with it? Which ones either provide input to the process or receive its output.

We talked before about authorization; the PDD lists what authorizations are needed for any systems or applications that interact with the process.

The PDD will also list the currently known risks. What could go wrong? What happens if something goes wrong? Does the process revert to manual performance until the problem is resolved? Who is responsible?

Lastly, the PDD will explicitly state the endpoints. How do we know that the process has completed?

A PDD can be accompanied by a flow diagram showing the current process and how that same process will/would look if automated. It can also specify the endpoints of a manual as well as an automated process.[2]

Technical Feasibility

There are several components to be validated. These include the following:

- Defined logics
- Rule-based steps
- Input and output data

---

[2] For more information, please see The Process Definition Document (robocorp.com).

- Complexity analysis
- Volume of transactions
- Development effort

The following steps are required to do the technical feasibility:

a) List all the applications the automation is going to interact with or use; this can be obtained from the process design document (see previous discussion).

   i). Determine if automation tool provides functions of scrapping or interaction with these applications.

   ii). Determine if there is any application that the automated process would not be able to interact with (e.g., any website that has complex JavaScript that updates its classes too frequently).

b) Will the automated process provide the required security (again, refer to the PDD)?

c) Can the process, when automated, run at the speeds and times required by the business?

d) Are there any regulatory requirements?

■ **Tip**   You can see how the PDD will be used extensively during design. It's vital to assure that the information it contains is complete.

Proof of Concept/Prototyping

In some cases, the RPA team may not have experience with the systems or applications with which the solution will interact. Or there might be very serious risks, such as privacy breach or risks to stability, that are involved. In these cases, having a proof of concept or developing a prototype can be highly beneficial. It may not even be necessary to create a prototype or proof of concept for the entire automation, but only for those parts that are particularly risky. Work with your SMEs and "risk and compliance" department to determine when this might be necessary.

Solution Design of Automation (What Will the BOT Look Like?)

We recommend using a layered design approach. The following steps will guide you:

1. Controller (process): This is the main controller that controls the Bot execution. It can be seen as the "supervisor" of other tasks within it.

2. Task container (business objects): This is the "nuts and bolts" of the Bot. Here are the various small tasks that will be used repeatedly by the Bot to execute the steps required. These steps are executed by the controller.

3. Communicator: Every automated process must issue some kinds of reports to show its results. Additionally, if transactions enter the Bot that it doesn't know how to handle, those transactions must be sent to someone for manual handling. As suggested by its name, the controller is used for these and other communication purposes as may be required.

■ **Pause and Consider**   What are the communication needs for the processes you are automating? How much information, if any, does senior management need? Where will transactions that cannot be handled by the Bot be sent? What record of them will be kept, and where will it be sent, stored, etc.? How often will a report of the process execution be created, and where will it be sent, stored, etc.?

Let's prepare the solution design document (SDD).

Hemant Joshi, CEI and Managing Director of eGleis Technologies, detailed what a SDD is.[3]

To summarize, a solution design document is a document that describes at a high level the design and the optimal way of implementing a technical solution to your project. It is created for every business process that is automated using RPA technology. This document is created by the RPA developer who will automate the business process, with input from the RPA solution architect who will review it before handover to RPA operations. It comes under the design phase of the RPA development life cycle.

---

[3]`www.linkedin.com/pulse/writing-solution-design-robotic-process-automation-project-joshi/`.

A typical and effective solution design document (SDD) contains the following components (see Appendix 8 for a template):

- High-level design: This will include an automation flow diagram and an infra architecture diagram.

- Information security: What are the compliance requirements? This can be obtained from the process design document (PDD).

- Prerequisites to automation execution: Consider hardware, software, licenses, etc.

- Exception handling: What will be done with transactions that enter the process, but cannot be handled by it? Every process has some "exceptions to the rule," and you must determine explicitly how these will be handled.

- Debugging tips: What suggestions can you include to help developers in the future to resolve problems? Remember, the developer who initially develops the automation may not be available if problems arise with it at some future date.

- Success criteria: Lastly, document what needs to happen for the automation to be successful.

Tip While each automation will be different, keep in mind that there may be reusable components. This will add to the efficiency of the entire RPA initiative.

Writing a best-fit solution design document (which is sometimes referred to as a solution architect document) for any RPA solution is the most critical piece in the RPA development process.

Tip Assuring a complete and thorough PDD (process design document) will greatly assist in the creation of a good SDD (solution design document).

The following are the major points to keep in mind in the process of writing a good RPA solution design document.

Development/Implementation

Once the SDD has been completed, reviewed, approved, and signed off, development of the Bot can begin.

In developing an RPA solution, development or implementation activities are different from those in more traditional settings. For a newly established RPA development team, the work can be divided into three phases. The first is solutioning the process steps. What will the automated solution exactly look like? What will the Bot do? This establishes the vision or view of what the technical solution will be.

Secondly, the Bot is designed in full detail. Using the solution that has been envisioned, the design is now created in detail.

Finally, the actual coding and implementation of the Bot in the automation tool used by the organization is done. This, of course, follows creating the vision of the technical solution, and then the actual work of coding is performed according to the design that has been created.

As with any development effort, coding standards must be followed. If your organization does not have coding standards, these can be developed by the team.

■ **Pause and Consider**   If your organization doesn't have coding standards, the introduction of an RPA initiative is a fine place to develop them. How these will be defined in your organization is up to you. But you need to assure that they are sufficiently rigorous and can be used across platforms and business units.

If your team is unsure of the importance of coding standards in software development, you might want to remind them of some compelling reasons. Coding standards help improve the quality of the overall software system. They reflect a harmonized style, as if a single developer wrote the code in one session. Coding standards also reflect more than just the software; they are a reflection on the organization itself. Additionally, they greatly help to improve the maintainability of code; any qualified developer following the established standards will be able to maintain the code. Having coding standards also helps to limit risks; there is less unknown because by adhering to the standards, anyone can read the code. Coding standards also improve code quality; all developers must adhere to a professional, pre-approved standard. And lastly, having these standards helps all developers be familiar with the code structure.

Nishant Goel and Satyendra Shinde of *BOTmantra* identify five (5) key coding standards and several subcategories. The following list shows them all:

Top 5:

1. Readability
    a. Name convention standards and compliance
    b. Zero usage of junk code
    c. Componentization
    d. Simplified logic
2. Configurability
    a. Performance parameters
    b. URLS, files, and folder paths
    c. Email IDs
    d. Credentials
    e. Business rule threshold parameters
    f. Log messages
    g. Email formats
    h. Generic design
3. Reliability
    a. Exception handling
    b. Best interaction technique
    c. Memory leakage avoidance
    d. Auto-recovery and auto-healing mechanisms
4. Security
    a. Authorization
    b. Authentication
    c. Credential management
    d. Business data storage
    e. Data sharing
5. Performance
    a. Delay management
    b. Parallel execution

    c. Interaction technique usage

    d. Memory management[4]

We will now look at each component in detail:

1. Readability: How easy is it to read the code? Do naming conventions exist? Have naming conventions been followed? Is the code as simple as it can be? Are there comments that explain modules or even separate lines of code, if necessary?

Tip The developer who creates the code may not always be available if problems arise. Any qualified programmer should be able to read and understand the code.

---

2. Configurability: Bots must be created with the possibility of further changes. Perhaps new parameters will be added, or different systems will need to "communicate" with the Bot. Are you including performance parameters, files paths, folder paths, various credentials and business rules, etc.?

Tip Always design with the future in mind.

---

3. Reliability: What degree of accuracy is required? Remember, we said that automating a manual process with RPA is best done when there are few events that must be removed from the process for manual handling. What exception rate is acceptable for the particular process being addressed? How will those exceptions be handled? Are there auto-recovery mechanisms in place to avoid downtime?

Tip It is sometimes believed that customer-facing processes require a higher degree of accuracy than internal processes. Don't cut corners on any process. Doing so will damage your credibility.

---

4. Security: A rigorous risk assessment will greatly increase security. As mentioned previously, the risk assessment will be ongoing: in the early stages, only limited information is available, so a complete view of all risks will be impossible to attain. That is fine! But as you progress with the project, additional risks will be identified. Each must be either mitigated or accepted, unless they are considered to be so great that the Governance Committee

---

[4]https://botmantra.com/rpa-coding-bestpractices/Accessed on April 15, 2021.

decides to cancel the project. Is it important to know: What authorizations are required to use, change, or stop the automation? How will the Bot "know" if it should run or stop? How will authentication be accomplished? Where will data going in and coming out of the process be stored?

■ **Tip**   As mentioned previously, early involvement of your risk and compliance team will be highly beneficial. Keep in mind that that team might have a different name in your organization.

---

5. Performance: How will you optimize performance? What interaction technique best suits this particular process? Can the Bot run on two or more processes simultaneously? What is the most efficient business logic configuration for this process?

■ **Pause and Consider**   The four previous components all impact performance. What are your performance standards? What is required by your organization? Defining these requirements clearly will greatly assist you in defining what is required for readability, configurability, reliability, and security.

At this point, you may be thinking that there is an awful lot to accomplish to introduce and implement a successful RPA program in your organization. Yes, there is a lot to do. But what you are doing is nothing short of revolutionizing the way your organization operates. Such an undertaking will take time and effort but, if done properly, will bring significant benefits. So don't be intimidated by the work required; follow the steps we have included. A task is never quite so daunting when taken in small steps.

■ **Pause and Consider**   As with each of these standards, you need to determine which ones are required for your organization and how rigorous each should be. What configurability components are vitally necessary within your organization? Which are "nice to haves"? How detailed should each be? The answers to these questions will vary from company to company and even within companies, from one business unit to another. Also, what is required for one automation may be more or less than will be required for another.

■ **Tip**   As the team gains more experience, technical feasibility will be folded into design activities, but for the present time, it should be separate.

Code Reviews

For many organizations, code reviews may be informally performed, if done at all. Establishing a code review practice in your development activity or team is very important. Dan Radigan, writing in *ATLASSIAN Agile Coach*, says that "Code review helps developers learn the code base, as well as help them learn new technologies and techniques that grow their skill sets."[5]

In the process of code reviews, other developers review the code to assure quality; remember, there is no place for "spaghetti code today". Code reviews help ensure the readability of the code.

The code review also ensures that coding standards, which we've already discussed, are strictly followed.

Another advantage of code reviews is that they help to assure that code can be easily maintained.

They also assure that exception handling standards for the particular process being automated are followed.

Code reviews, like all aspects of development, are more efficient and effective when certain guidelines exist. Having clearly documented coding standards is key. Also, it is helpful to provide and follow checklists; create these from your code review standards.

Additionally, it is often beneficial to provide pre-review training, to assure that developers are familiar with the standards they must follow.

Having code templates is also very helpful.

■ **Tip** If you reduce code reviews to an afterthought, something you will do "if time allows," be prepared to fail. Code reviews are an integral part of any RPA project.

Testing

Once code reviews are completed, and any changes to code have been made and re-reviewed, you are ready to test.

For successful testing, the testing teams must be completely familiar with the process. Knowing and understanding the relevant PDD (process design documents) and SDD (solution design documents) are key to success.

---

[5] www.atlassian.com/agile/software-development/code-reviews#:~:text=Code%20review%20helps%20developers%20learn,that%20grow%20their%20skill%20sets.

An exhaustive list of test use cases must be created; this could be an overall test plan. This will include testing data for the automation.

■ **Tip** As you progress with the RPA initiative, you will learn from defects and thus improve testing procedures. Even the most comprehensive test plans will not be bulletproof; there will be continued fine-tuning of testing procedures.

Testing automation can be a challenge for several reasons. A major one is that many organizations don't have testing environments. Also, it is sometimes difficult to get data, and the Bot may behave differently based on the data it is given.

Additionally, the more applications with which the Bot needs to interact, the more difficult it becomes to build and manage parallel environments for testing. Because updates to the applications happen during Bot execution, sometimes testing can only be done in the production environment. This includes the Bot needing to update some screens that may have wrong data or the possibility of it making incorrect or partial updates.

Following established testing techniques is vital when transitioning to RPA. We will look at some of the aspects of RPA testing that you may find different from testing in other methodologies.

Also, please remember that the people performing testing must be trained. Someone with minimal training may be able to execute the testing (see step 4), but steps 1, 2, and 3 require a degree of knowledge that if learned "on the job" carries high risks. Be sure that the people performing these steps are trained in doing them. If a less experienced person is going to perform step 4, be sure they are trained on what's required for step 5.

1. Know the business process: Those performing the testing must know and thoroughly understand the business process that is being automated. Without this knowledge, they will be unable to properly develop and implement effective tests.

   The best way to understand the process is to review the process definition document (PDD), the solution design document (SDD), and any other documentation that was created during the design phase of the automation. Once the team understands the business process, it can move on to the next step: creating the test scenarios to actually test the codes.

■ **Tip** If something is unclear, don't hesitate to contact the business owner for clarification. Do not cut corners in understanding the process.

2. Create test scenarios: This is another key area. Review the PDD and the SDD, with special attention paid to the SDD which, if done properly, will list the most important scenarios to be tested. But don't ignore the PDD; very useful information for building test scenarios can be gleaned from that document, too.

3. Write test scripts: Now that the testers thoroughly understand the process to be automated and have created test scenarios, the next step is to create a series of test cases. Test cases are very specific: they list the inputs, expected output (or outputs), and a column (often test scripts are written in a spreadsheet) to indicate if the test passed or failed and another column for any notes. This column is especially useful for tests that fail.

■ **Tip** To maximize the effectiveness of test scripts, they should be reviewed by the design team. This "second set of eyes" is very familiar with the process and can assist in finding any testing gaps.

4. Perform testing: Now that you have completed all the preliminary work, the actual testing is performed. Complete all the work of each test case, documenting if each passed or failed, with any relevant comments.

5. Rectify defects: Do not expect all test cases to succeed; if they do, it's likely you have not created sufficiently effective test cases. Defects are common, and identifying them is a main purpose of testing. All failed test cases should be fully documented; too much information is better than too little. There must be sufficient information for the development team to be able to correct the errors.

■ **Pause and Consider** What is the testing process in your organization today? How must it be adjusted for RPA testing? Will there be resistance to these changes? If so, how will you overcome it?

The test documentation must be stored in a repository that is easily accessible to the people needing access to it. Remember that the developers who will need to see it are probably working on multiple projects. You need to be sure they can quickly and easily find the information they need.

■ **Pause and Consider**   Where is test documentation currently stored in your organization? Is it sufficiently accessible to suit the needs of your new RPA initiative? If not, what changes need to be made? Who will make them?

Much of what has been discussed in this session relates to quality assurance (QA) and quality control (QC). That is not coincidental! Let's look at the definitions of each:

Quality assurance: "Part of *quality management* focused on providing confidence that *quality requirements* will be fulfilled."[6]

Quality control: "Part of *quality management* focused on fulfilling *quality requirements*."[7]

So quality control can be seen as fulfilling the needs of quality assurance.

■ **Pause and Consider**   Does your organization have formal quality assurance and quality control functions? If so, how can they best be integrated into your new RPA initiative? What, if anything, will need to be done differently? If there are no formal quality assurance and quality control functions, how can you still assure that the needs of both are met within your RPA program?

Automating Test Execution

When the RPA team is automating customer or client processes, it will be beneficial to automate its own processes also. This includes automating the testing process of Bots.

Since all Bots are unique in their function, before deciding to automate their testing, there are a few things that need to be considered. Determine how frequently will testing automation be used. For each particular Bot, how simple or how complex would it be to create automation testing. Also, determine if the testing automation you may create could be used to test other Bots in the future; try to determine its reusability.

If it is determined that automated Bot testing would be beneficial, there will be three components to do so. You will first do automation to set up the

---

[6] https://asq.org/quality-resources/quality-assurance-vs-control.
[7] Ibid.

environment and create the required test data. You will then execute the Bot automation. Finally, you will gather the output data after the Bot execution and validate the output to create a test summary.

Case Study

Now let's look at how solution design is implemented in our case study which was introduced in Chapter 4. We moved forward with it in Chapter 5. Now we will see how our hypothetical case would move through solution design.

Let's start by creating the SDD. In this document, we need to populate a variety of information. There are a variety of templates available online; use the one we provided or any other template that seems to meet your needs.

Solution Design Document

Phone and Address Change

		Comments
Author	John Williams, CSS Supervisor	
Date	July 28, 20xx	
Version	Draft 0.1	Initial draft

Approval	Signature
Mary Brown, Manager, Customer Service	

Purpose of This Document

This document contains the solution design for the phone and address update process. It presents the high-level "As Is" process steps and the "To Be" process steps that will be automated.

The current process is as follows:

- The request to change an address and/or phone number is received in the specified mailbox.

- The CSS verifies the client ID and updates the information as required.

- The CSS confirms the update with the client.

- The CSS maintains a record of updated information.

Flow Diagram

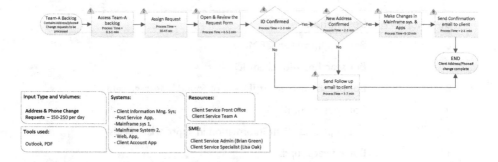

Goal of the Automation

The goal is to create a Bot that will retrieve the emails, read and parse them, update the information as specified, and inform the client that the information has been updated.

Applications Involved

Name	Internal/ External	Type	Credentials Required?	Read/ Write
Web Application – Client CRM	Internal	Network-hosted application	Yes	Both
Mainframe 1	Internal	Mainframe	Yes	Both

Inputs

Name	Description	Initial Value
AddressUpdate_ ClientID	An email that will contain the client ID and a new phone number and/or address	Address and/or phone information

Process Steps

First Sub-process

Step Number	Description
1	Bot starts.
2	Bot logs into email account.
3	Bot reads emails.
4	Bot extracts client information.
5	When all emails have been read, Bot sends completion message.

Second Sub-process

Step Number	Description
6	Bot logs into Web CRM.
7	Bot searches with client ID.
8	Bot doesn't find client ID – "not found" message sent.
9	Bot finds client ID.
10	Bot populates client information in appropriate fields.
11	Bot activates update.
12	Bot sends completion message.

Third Sub-process

Step Number	Description
13	Bot logs into Mainframe.
14	Bot searches with client ID.
15	Bot doesn't find client ID – "not found" message sent
16	Bot finds client ID.
17	Bot populates client information in appropriate fields.
18	Bot activates update.
19	Bot sends completion message.

Fourth Sub-process

Step Number	Description
20	Bot sends email to Client.

Exceptions

- Business

  - Client ID is missing in the email: Send exception message EmailMissingClientID.

  - Last name or first name is missing in the email: Send exception message EmailMissingFirstOrLastName.

- Technical
  - Web CRM URL not reachable: Send exception message CRMInReachable.
  - Login password failed – Web CRM: Send exception message LoginFailedPWDExpired.

Success Criteria

The process is complete when the Bot sends a job completion notification daily at (time – TBD).

Debugging Tips   If the Bot fails to log in to email, run the email subtask. If there is an error in updating the address or phone number in the web application, run the web application series of tasks, and check if updates are done successfully by the Bot. The same procedure can be followed to check to assure that the mainframe is updated.

Maintenance

This Bot has normal priority. If Bot fails or errored out, run the database script to reset the Bot and restart the Bot.

Bot Scheduling

Bot will run every day, Monday through Friday, at 11:00 in the morning.

Disaster Recovery

In case of complete failure of the Bot, the team will need to revert to the manual process:

- The operations team will perform the process manually, until notified that the issues have been resolved.

- The RPA team will be responsible for operational tracking, monitoring, and maintenance. In case of any temporary outages or planned downtime, the COE will communicate to business well in advance and expedite the restoration of services.

- Process owners are expected to systematically review and audit the results to ensure the functional value is achieved.

- Any changes to host systems will be proactively communicated to members of the RPA team far enough in advance to assure they can make any required adjustments.

- The process owner will receive notification emails for all failures.

Glossary of Terms

Term	Description
RPA	Robotics process automation
SIT	System integration testing
SLA	Service-level agreement
SME	Subject matter expert
UAT	User acceptance testing

Appendices

Appendix A: Exception messages

- EmailMissingClientID
- EmailMissingFirstOrLastName

Let's look at these components in detail.

The cover page includes several components: name of document. This is the SDD, and it is necessary to specify what project it pertains to. The document author is the person responsible for creating the document. That may be the architect or designer, but it could also be another member of the team. The document version should indicate whether it is the first or a later draft, if it is final, and if it has been approved. Generally, draft versions start with 0 and are incremented as the draft is reviewed and updated: 0.1 is the initial draft, 0.2 would be a draft that has been updated, etc. The "Comments" column would be used to make any explanations. For example, Draft 0.2 might have a comment such as this: "Revised following review with Mary Brown, who requested that XYZ be altered."

Also on the cover page is the space for the approval. The Name/Role is the name and role of the person who is responsible for approving the document. That could be the product owner, business owner, delivery manager, or another role. The "Signature" would be electronic when the document is approved by the person specified.

■ **Pause and Consider** Remember that versions (draft, final, etc.) and role names (product owner, business owner, etc.) differ from one company to another. Use the names that pertain to your organization.

The "Purpose of This Document" provides a very short description of process automation overview. This is very helpful for understanding what the Bot is going to at a high level; additional detail is included in later parts of the document.

The "Flow Diagram," which was developed in the PDD, is included here, to provide developers with an "at-a-glance" understanding of the current manual process that they will be automating.

The "Goal of the Automation" starts to drill down to state more specifically what the Bot will do.

In "Applications Involved," you will list which applications the automation is going to interact with. By identifying all applications, you can discover what kind of access the automation will need to interact with each of them.

The developers need to know what the "Inputs" are. In our example, there is only one input (email from the client), but this is just a simple case study. Your "real-world" work may include more.

While a very high-level description of the process has been provided, that is not something that the developers can build from. Now you need to list the individual "Process Steps" in detail. In our case study, there are four, simple sub-processes. As you can see, they are very specific and at a very low level.

Any process is going to have exceptions. In our case study, that could be an email that doesn't include the client ID, or the client ID may have too few or too many numbers; the new telephone number could be missing a number, etc. These "Exceptions" must be dealt with by the Bot, and this section explains how that will be done.

The Bot has a beginning point and an endpoint, and it's necessary to assure that anyone monitoring it, either in real time or periodically, knows that it has completed its work for each cycle (hourly, daily, etc., as specified). This "Success Criteria" must be defined. Will it be an email to a manager, sent by the Bot, to say that all emails in the specified mailbox on July 29, 20xx, have been processed? Will it be a notation on a larger report or dashboard? Whatever it is, this is the place to specify it.

No application will ever run perfectly all the time. Because of that, the SDD should include some basic "Debugging Tips." These are suggestions you can include to help developers in the future to resolve problems. Remember, the developer who initially develops the automation may not be available if problems arise with it at some future date.

Maintenance

Once the Bot is deployed into the production environment and is live, it starts following its schedule. Then maintenance comes into picture. Every Bot has an SLA (service-level agreement) to provide support that sets expectations of the business in case the Bot fails to perform as required. Mainly, there are two aspects to it. One aspect is technical and the other is related to the business side. Technical aspects cover when the Bot is unable to perform because of any technical reasons, for example, servers are down, infrastructure maintenance, etc. Business issues might be related to the Bot being unable to perform as per requirements because data is not available or when input systems that feed the Bot are not ready at the time of Bot execution.

In case of technical reasons, the main responsibility of the technical team or COE is to communicate to the business information about the outage and also share the situation with any other relevant stakeholders. They should all be kept apprised of the progress and expected time when Bot will be available as per schedule. For any business reasons, the technical team needs to coordinate with the business team so that the Bot can be stopped or suspended until the business reasons are resolved; the technical team will also work closely with the business team and/or COE team to keep everyone on the same page.

Scheduling: The developers need to know when the Bot is to run. This may be daily at a certain time or multiple times; it might be weekly, monthly, etc. This will have been determined early on in discussions with the business and will be documented here for the development team.

Disaster Recovery: What happens if the Bot ceases to work? Either it stops work altogether or is malfunctioning to the point that it isn't doing its job. Again, what happens in this case will have been discussed with the business much earlier, usually when initially creating the risk assessment. Whatever the determination is should be included here. At a minimum, reverting to the manual process until the Bot has been repaired is one aspect of disaster recovery.

After completing SDD, it can be sent for sign-off as per the process in your organization.

Once SDD approval has been received, you are ready for developing the Bot in your selected automation tool. The steps included in this book will be effective with whatever tool you use.

Testing

After building the Bot to automate the process in our case study, the next vital activity is testing the Bot.

To capture all tests, testing execution, and its results, a test summary document can be created. Keep the test summary document as simple as possible so it will be easy to understand. Here in our hypothetical automation, we are going to follow a test summary that has three tabs:

- The first tab contains information about the Bot and test results.

- The second tab has test scenarios with results.

- The third tab has test data.

Test Summary

VERSION HISTORY			
Author	Version	Date	Comments
Allan Brent	0.1	03-Jan-21	Initial version

APPROVAL DETAILS	
ID	Automation - 001
Business Lead	John Pattorson
Business Sponsor	Alex Macleod
Approval Date	
Approval Evidence lin	N/A

TEST RESULTS	
Number of Test Cases	3
Test Cases Passed	3
Test Cases Failed	0
Test Cases Blocked	0

ENVIRONMENT DETAILS				
Version	Environment	Environment Description	Risk	Risk Mitigation
1	Test	Testing environment for the bot	No	N/A

* If tested in Production, then Risk and Risk Mitigation should be stated and signed-off by the Business Sponsor prior to the testing

Note that this tab has a Version History, Approval Details, Test Results and Environment Details. The information required in them is self-explanatory. For "Environment Details," include any risks that pertain to testing in that environment. This is especially critical if testing in a production environment.

Test Scenarios and Results

Prerequisites	Process Step Number	Test Case Description	Expected Result	Actual Result	Test Case Actual Status
Email account is set up on email server.	I	Bot logs in to email inbox.	Bot successful login to email account.	Bot able to log in	passed
Web CRM application is up and running.	2	Bot logs in to Web CRM and performs update to one address and phone number.	Bot successfully updated one record.	Bot able to update record	passed
Mainframe I is up and running.	3	Bot logs in to Mainframe and performs update to one address and phone number.	Bot successfully updated one record.	Bot able to update record	passed

This section is also self-explanatory.

■ **Pause and Consider**   You will note that in keeping processes and documentation as simple as possible, you will have more success with your RPA initiative. But remember not to compromise quality as you simplify.

Test Cases

ID	FRD-Based Scenarios
Auto-001	Bot successfully logs in to email inbox and reads emails.
Auto-002	Bot successfully logs in to Web CRM and updates address and phone number.
Auto-003	Bot successfully logged in to Mainframe I and updated address and phone number.

This section shows the specific test cases that must be executed.

(For the purposes of the case study, we are only showing user acceptance testing).

Following testing, business sign-off is required. This is done by forwarding this testing document to the business owner and receiving and then storing their approval, usually as an email.

Testing can be completed in iterations; it can also be done more than once to assure that the Bot delivers what is expected and is able to handle exception situations as per requirements.

At this point, the Governance Committee is advised of business approval and solution design; no meeting is required. If the committee members have any questions, they will notify the RPA team.

After completing testing and obtaining approval, preparation for packaging and deployment can start; this is explained in detail in the next chapter.

Solution Deployment, Maintenance, and Retirement

Once the automation/technical team completes the development and testing of the Bot in the UAT/testing environment, it's time to deploy it in the production environment. Make sure your technical team is following all the procedures already in place for testing and development of the Bot so that a quality solution that meets the expectations of the business goes into production. To move or deploy the Bot in production, a detailed and accurate procedure should be followed to avoid any unexpected issues while deploying. In this chapter, we will go through in detail everything you need to know about the deployment of the Bot to production, how to manage and support it once it's in production, and how to retire the Bot whenever that is required. This will prevent any unpleasant surprises during the whole live life cycle of the Bot.

© Robert Fantina, Andriy Storozhuk, Kamal Goyal 2022
R. Fantina et al., *Introducing Robotic Process Automation to Your Organization*,
https://doi.org/10.1007/978-1-4842-7416-3_7

In addition to these topics, we will also discuss auditing and performance improvements of the Bot.

Solution (Automation/Bot) Deployment

Now is the time to operationalize the work you have done. You have received approval at each gate; updated risks, ROI, size, etc.; and designed and created the solution. Now you are ready to put the Bot in production.

Some larger organizations have a Center of Excellence; many do not. The term COE means different things in different companies, and the functions may be performed by the COE or the technical team. So when we refer to the COE, please remember that it may have a different name in your company or may not exist at all.

This organization, whether a COE, another group in your company, or the technical team, will have ultimate responsibility for deploying the new Bot. It will assure that the scheduled date can be met and that any associated applications will not be compromised by the new Bot and will monitor performance with the RPA team.

For any RPA program, there are many functions required; remember, they may be performed by a Center of Excellence or similar team, or by the technical team. You must consider the architecture of the robotic operating environment (infrastructure support, technology choice). Also, RPA operations (maintenance, support, monitoring, training of new resources, change management) are vital to know. What will your governance model look like? This includes integration to overall organizational structure, oversight of resources responsible to support all functionalities of the RPA program, compliance to policies and procedures, security system access, process prioritization, and the escalation path.

You need to know the delivery (process discovery and assessment, solution design, testing, deployment, development of standards).

In addition, the COE, if there is one in your organization, among its other responsibilities, maintains all the common functions required for RPA for the various lines of business. This makes management of the Bots simpler and more cost-effective. The COE also establishes standards and best practices and fosters continuous improvement.

■ **Pause and Consider**   We have said a lot about the Center of Excellence and acknowledged that there may not be one in your organization. We have also detailed, and will continue to discuss, the role of the COE. Who in your organization would be responsible for these tasks? Remember, it might not be one group, but the work may be distributed over several groups or individuals.

As you can see, early involvement with the COE will greatly assist you in deploying Bots on time and effectively.

The COE is also vital in managing access management and security. The COE assures that the Bot only has access to the resources (applications, systems, etc.) that are required for that particular automation. User IDs and passwords expire periodically, and the COE assists in renewing them as required. This differs from other application management functions, and it's important to be aware of these differences when rolling out your RPA initiative.

Because the RPA program in your organization will evolve quickly, the COE enables the build of the new infrastructure and the adoption of new automation tools.

Many organizations, however, do not have a formal COE. If that is the case for you, the same functions must still be accomplished, but they will be performed by the technical team.

■ **Pause and Consider** When we say "technical team," we refer to the broader team beyond just developers. This can include such roles as process analyst, architect, business analyst, infrastructure analyst, developers, asset management, security, etc. What roles comprise the technical team in your new RPA program?

The technical team will work closely with the business partner who has requested the automation. The same tasks — establishing standards and best practices, continuous improvement, security, access management, etc. — must still be performed, but the technical team will take the lead on each one.

There are both advantages and disadvantages to having a COE, so you shouldn't feel at a disadvantage if you don't have one.

When there is no COE or equivalent group in the organization, the technical team takes over all responsibilities of COE. Less communication is required as it will only be between technical team members now. Faster action and quicker response times can be achieved. The technical team makes most of the decisions, and support and maintenance needs less management since there is no additional communication and collaboration required with any external team or department. The technical team can establish processes according to their own specific needs and improve or modify them in a timely manner.

Where there is a COE, it must follow organization guidelines and generalize the processes and standards. These may be less suited to individual teams, and they may need to be modified or updated to accommodate the requirements of a particular Bot/team/department. So while there are certainly advantages to having a COE, there can be drawbacks.

■ **Tip** If your organization has a COE, do not attempt to bypass it, even if you think it will expedite deployment. This will cause a wide variety of problems and could potentially sink your RPA program.

I. **Process setup for deployment and release.**

a. Prerequisites for deployment: Sizing of the Bot(s), etc.

Before deploying any Bot, there is a wide variety of information that must be determined and shared with various parties. These include information about

- The size of the automation. It could be the number of files, any database scripts, or other settings like scheduling, frequency, and number of Bots required. Sizing can be done on the basis of many factors. It could be based on the number of Bots needed to complete the work in the specified time, for example, if work can be completed in two hours, only one may be required, but if the same work needs to be completed in thirty minutes, then four or more Bots may be required. The size may depend upon the complexity of the automation or the steps that need to be automated, if they are complex in logic or must interact with multiple applications at the same time; this adds complexity to the automation, so more control over the Bot is required.

- Size can be categorized as small, medium, or large. This categorization can be determined by many factors. Preliminary understanding will be obtained in the earlier phases. As you gain more information about the process, you will define the size.

- The factors that determine size include the number of process steps being automated, number of systems accessed, volume of work, etc. Note that at this point, you know if it is "small," "medium," or "large." As you have progressed through the process and gained more information, you now are able to definitely say the automation's size. During "opportunity assessment," you will determine a high-level understanding of sizing, but true size will depend on solution and scope that would be finalized at solution architecture.

- Remember, don't be too concerned about getting the sizing exactly right. These are general terms.

■ **Pause and Consider**   For each organization, small, medium, and large will depend on the factors included herein and your own organization. What would constitute "small," "medium," or "large" in your organization?

In addition, the Bot could be attended or unattended. "Attended" means that someone is watching the execution of the Bot and observing it while it is running. This is done so action can be taken should the Bot error out. An attended Bot is always attended.

An "unattended" Bot will still be observed during deployment, but once the operational aspect is shown to be successful, it will no longer be "attended."

- Another piece of information is the scheduling of the Bot. How many times will the automation run each day, week, month, or even year? Or will it be "on demand"? This information was determined early on, when the purpose of the request to automate was initially received.

- The volume of work that the Bot is expected to handle is also vital. If the work is stable and doesn't change the period of time that the Bot will run, it will be simpler, but if volume may change significantly from run to run and there is no information or pattern or historical data about the work, then this must be carefully documented. Having historical information (e.g., the volume is usually doubled during the first three days of the month) will greatly assist in deploying a successful Bot.

- It is necessary to verify the access to be used by the Bot in production; this includes the environment, user IDs, application access, and network access.

- Verify that the emails are correct for communication. For example, if the Bot is to send those exceptions (situations that enter the process that, for any of a variety of pre-identified reasons, cannot be handled by the Bot) to a specific email, assure that the address is correct.

- If the Bot is an update or enhancement to an existing automation, you must plan for disaster recovery. In the event that something goes wrong with the update or enhancement, you need to be able to quickly return to the Bot prior to the update or enhancement, so at least the work that had been processed can continue to be.

- Any software version has three parts:

  - Release number (sometimes referred to as "major version number"): For a new automation, the release number will be one (1). Release or major version numbers also change whenever there is some significant change being introduced. For example, a large or potentially backward-incompatible change to a software package.

  - For enhancements, including adding steps of a manual process into the automation (remember, the initial automation may only have automated some of the manual steps; at some point, additional steps may be automated using the same Bot), the release number would be incremented by one.

  - Minor version number: Minor version numbers change when a set of smaller features is rolled out within the Bot.

  - Patch number: Patch numbers change when a new build of an existing Bot is deployed. This is normally for small bug fixes or updates to templates, etc., used by the Bot.

  - An example of a number would be: 1.002.013. This indicates that it's the original release (or major version), there have been 2 minor updates, and 13 patch changes (bug fixes, template updates, etc.).

- Identify the specific people (not just the roles) in the business, and the SMEs, who must be present at the time of deployment.

Pause and Consider   Throughout this book, we have specified identifying roles, not names of individuals. Now we are specifying individual names. This is because at this point, you need to know exactly who will fulfill these responsibilities.

- If your organization has a COE, determine the specific person or people from the COE who will be present during deployment. This should not be at all difficult to determine, because if you do have a COE, you have been interacting with that group from early in the process.

- Validate that all project and/or folder structures are consistent with all guidelines on production automation tool being used.

- Assure that all applications and systems are accessible from the production environment (URL, network, etc.) by the Bot user ID and any other user ID used for the application itself.

- Ensure that the infrastructure is capable of adding new automation and Bots according to the schedule.

- What are the risks? At this point, all identified risks must have a mitigation plan, with a role assigned to it. The mitigation plan may be simply accepting the risk, but there needs to be a role assigned, so that if that risk event happens, that person can be contacted to see how they want to handle it.

b. Instructions and documentation for deployment: As indicated by the paragraph and lists earlier, there is a lot to do to ensure a smooth and successful deployment. In order to accomplish this, there need to be detailed instructions documented for the specific deployment. While the items in the bullet list mentioned previously must all be achieved, each automation will have tasks specific to it. These include preparing a transition to production form and sharing it with COE if there is one or with the pertinent members of the technical team if there isn't. Please refer to transition to deployment form.

c. Release options (pilot release): When the automation or the Bot is large and complex in nature, and is business critical, then you can choose to deploy the Bot in a constrained manner instead of going full live. There are many options that can be tried or adapted. This technique is also called pilot release.

There are various pilot release options. These include

— Single Bot only: If the Bot has to release with multiple Bots, for pilot release, the Bot can be single so that if errors occur, they can be handled properly. Sometimes if the full release of the Bot is single, then restricted mode can be adopted.

— Restricted mode: In restricted mode, very light work is assigned to the Bot in the first few days, and performance is watched.

— Ghost mode: When the work done by the Bot is visible to end users or has a big impact on the business (e.g., there are dependent services on the work or updates done by the Bot), it is preferable for it to be validated prior to being delivered to the end user during the warranty period (more about warranties later). This can be done by diverting the output to some other area of the system or another environment (email to a member of the technical team, creation of a report, etc.). If the outcome is what is expected, the Bot is rerun in the production system but not in ghost mode.

There are no hard and fast rules; it all depends upon the situation and how the outcome of the Bot is to be captured. All these factors must be considered prior to a pilot release so they can be effectively managed.

d. Negotiate deployment/release dates with the business: Often, the person requesting the automation wants it yesterday! During the initial discussions (see Chapters 4 and 5), you and the business owner and SMEs have come to a mutual understanding about a realistic date. This is seldom one specific date and is more likely a range of dates. For example, the business, in conjunction with the development team, decides that a reasonable date for deployment is between May 15 and June 8, 20xx.

This information would have been conveyed early on to the COE, if your organization has one. If not, your regular contacts with the business and with people responsible for the systems and applications you must impact would have been kept informed about the date range, assuring with them that it was reasonable.

Now it is time to finalize the dates with the business. You have far more information now than you did in earlier phases. You know the upstream and downstream systems and applications to be impacted. You know what needs to be done for deployment, and now you can use this information to nail down a specific date with the business owner.

If your organization has a COE, that group has been a third partner with the technical team and the business and now must be included in determining the release date. The COE would have been advised of the range acceptable to the business and technical team as early as it was known. If, at that time, the COE said that no date within that range was possible, you would have gone back to the business to renegotiate. Now it is time to work with the COE and the business to determine the date of deployment.

e. Warranty: When the organization has a COE, the warranty time is provided by the COE after deploying the Bot in production. In organizations without a COE, the warranty is handled by the technical team. Usually, the warranty is set for two weeks, but the dates of the warranty period are flexible. For very simple Bots, it could be only for one week, but decreasing the warranty period is the decision of the business and technical teams. Similarly, for very complex Bots, the warranty can also be increased to three to four weeks. Again, increasing the warranty is not something that happens very frequently, but is agreed to among the business, technical team, and COE (if there is one). This depends partly upon the resources required to provide service during the extra period and their availability.

It is also advised that in these situations, if there is a COE, it should be made aware of the extension as early as possible, so that they have sufficient time to be prepared.

Major Steps for Deployment

As you prepare for deployment, you must assure that the scheduled date for deployment has been agreed to by all stakeholders. Also, be sure that all required documents are signed off for the deployment process. This includes all documents required by compliance: requirements documents, code review, and test summary. Stakeholders include the release or delivery manager and the business product owner. There may be other stakeholders in your organization whose sign-off is required.

You will then package all the code files and create database scripts or any Bot scripts (depends on the automation tool being used) and any and all Bot documents (e.g., Read Me files).

■ **Tip** Some software tools allow the user interface to deploy Bots to production environments from stage or UAT environments. Check if the automation tool used in your organization provides the user interface for moving files from the UAT environment to the production environment.

You will need to share all required files (all those files created in the step mentioned earlier) and pertinent documentation with the deployment team. This could be through network share, email, or the automation tool itself.

■ **Tip** Some software automation tools may offer to build scripts, and those scripts may be invoked manually or via other means to deploy Bots. Again, you need to check if the automation tool that you are using provides the user interface for moving files from the UAT environment to the production environment.

As you approach deployment, schedule a meeting with the deployment team members to block their time for the time agreed upon for deployment. You need to assure that they are available for the deployment.

You will now deploy all files to production and run scripts on the database if required. Once all files are shared with the deployment team, all files will be copied to the production environment, during the time scheduled for deployment. Validate that all required scripts have been run.

■ **Tip** Bots can also be deployed by physically copying code files to the production environment and changing the required databases as per requirements. As stated previously, you need to determine if the automation tool that you are using provides the user interface for moving files from UAT environment to production environment.

You will schedule the Bot as per requirements (e.g., hourly, daily, weekly, working days only, etc. This was decided earlier in the process). Then you will test run the Bot in production if required. This assures that the Bot is working properly in production, because the environment is new. The production environment may be different from the test environment. After that, you will validate the results to assure the Bot is working properly and then announce the successful deployment to the business team.

■ **Pause and Consider**   Deployment steps vary with the software automation tool that is being used for implementing RPA in your organization. What changes will be required in development in your organization, depending on the tool you select?

The technical team will continue to monitor the Bot for a few days as agreed to with the business (and COE if there is one). This is for new Bots. For upgrades or other changes, the technical team will only monitor for one run, or one day, depending on what has been agreed upon.

We have provided a long list of items to do prior to deployment. How will you remember to do them all? The best way to avoid having anything fall through the cracks is to use a checklist. The following is a sample (this is also included in Appendix 11 for your convenience).

Deployment and Release Setup Checklist

Item	Item Details	Comments	Status*
Product (Bot) Name		Provide the name used to refer to the solution/automation. This name will be used in inventory, reports, change requests, etc.	
Deployment date and time if applicable, agreed upon		As per project plan, provide the planned/preferred date for the deployment.	
Size of automation		It could be small, medium, or large.	
Bot schedule		Specify the preferred schedule (times and days of the week). The acceptable range of start times for the Bot. The approximate runtime of Bot. Are there any service-level agreements associated with the process that affect the scheduling of the automation?	
Contacts during warranty		During warranty, the deployment team may need to contact a software developer on the delivery team that is familiar with the code. Ensure that the developer will be available for five business days after the deployment date.	

(continued)

Item	Item Details	Comments	Status*
Statement of segregation of duties		Delivery team will note names of software developer(s) who wrote the code.	
		Automation services will note names of who user acceptance tested the code and who on automation services is deploying the code.	
		To ensure segregation duties, automation services will ensure that	
		The software developer who wrote the code was not the same person who performed user acceptance testing.	
		The deployer of the code to production is not the same person who wrote the code.	
Volume of work		What is the expected number of items to be processed by the Bot in a designated time period (daily, weekly, etc.)?	
Access verification		Mention what type access is required and has it been verified in production environment.	
Email addresses for communication		Emails of business and technical contacts those will be contacted for any information, errors, or other reasons.	
Disaster recovery		Backout steps, post-implementation steps.	
Major/minor version		Specify if this is a small change to an existing Bot or a large, significant change.	
People who will be present at deployment		List the names and roles of the people who must be present at the time of deployment.	
COE contact		List the name and title of this person.	
Folder structures		List any new folders required.	
System/application accessibility		Document any known issues.	
Infrastructure capability		Document any known issues.	
Risk identification and roles responsible		Summarize this from the risk assessment.	

Item	Item Details	Comments	Status*
Deployment instructions/steps		Provide a detailed playbook for the deployment.	
Task Deploy Steps		Typically the automation services deployment manager specifies the implementation and backup steps.	
Pre-implementation steps			
Implementation steps			
Test the implementation steps			
Backout steps			
Post-implementation steps			
Communication channels established		Specify them.	
Production Support Model		Are you retaining break-fix responsibilities for this automation? If yes, provide remedy support queue name and email contact for the business technical support team. If no, automation services will engage the development team as part of the D2P process to ensure a smooth transition to maintenance knowledge transfer.	

"Status" could be "complete," "incomplete," or "not applicable." For "incomplete" or "not applicable," be sure to include a brief, explanatory comment.

Maintaining Communication Between COE/LOB and Dev. Teams for Deployment

We have mentioned that your organization may or may not have a COE. If there is one, it is vital that you work with the COE throughout the process of the automation initiative. There must be efficient communication about coding standards, code review standards, upgrading the automation tool, and deployment schedule. It is the responsibility of the technical team to set expectations. What is the date (range) that the business wants deployment to fall within? Will any infrastructure upgrades be required? What development environment will be used?

The COE, if there is one, defines who will maintain the development environment and who will maintain the Bot. Some might be maintained by the COE, some by the development team. If there is a COE, it maintains the execution of the Bot.

The technical team has the contact with the business, so the technical team has a greater role in collaboration and communication during maintenance. If the Bot didn't run, the business will contact the technical team, which will then, in collaboration with the business and the COE (if there is one), determine the next steps. This could be reverting back to the manual process until the Bot is repaired, rerunning the Bot once work is available, or stopping the Bot and restarting it when the business determines it can be restarted. Rerunning the Bot depends on the SLA; rerunning options are usually decided as part of deployment planning. We will look at some examples:

1. Stopping the Bot: The Bot could be stopped and the process performed manually, until the repair is made. Then the Bot begins again with the new, incoming information.

2. Rerunning the Bot: Perhaps the Bot didn't run overnight as scheduled. In this case, it could simply be run during the following day with someone monitoring it to assure that it runs.

3. Stop and start: If the Bot isn't performing as expected, and it is due to the load that the Bot is processing (needs some tweaks or updates), or if the manual process needs some changes, then the Bot can be stopped and the necessary changes made. For example, Bot is processing Jira cards (they are the input for the Bot), and the manual process steps to be executed before the Bot begins have not been completed (e.g., Jira cards have not been updated or otherwise set up per the Bot's requirements), the Bot can be stopped until the required work in the manual process is complete. Then the Bot can be restarted.

Level 1 support is done by the COE: changing the schedule, changing the frequency of run, changing the number of Bots running at one time, updating user IDs and passwords, changing the priority of the Bot, and including/ excluding any email IDs that need to be contacted in case of information from the Bot. All changes and updates which are not impacting the actual running of the Bot from the technical side. Example: Bot is running fine, but some changes in the settings of the Bot are required. These are nonfunctional requirements.

Level 2 support is done by the technical team: when the Bot is not performing as per requirements/expectations. This pertains to functional requirements.

In the absence of a COE, both Level 1 and Level 2 support are done by the technical team in collaboration with the business owner.

Maintenance of Bots

3.1 Setup and manage the support queue.

a. Agile: Dev. team rotates the term (weekly?) for on-call support and will respond to COE/LOB contact about malfunctions. An inbox may be established to report on Bot performance and exceptions. This is monitored by the Dev. team member designated for the particular week. This allows all the team members to be familiar with all the Bots. In a more traditional methodology, one or two members of the technical team are assigned to the support queue, although a more rotational support may be implemented to increase team members' knowledge of all the Bots.

While setting up the support queue, steps between the COE and the business should be very clear. This includes when an incident is reported to the technical team, it is determined: should the Bot be stopped; who is the contact on the business side for that particular Bot; can the Bot be rerun (decided in advance that it will be rerun every time, or will that be an individual decision); will it be stopped immediately every time; etc.? For some Bots, it is known that the business wants it to run again if it fails.

A short description of "next steps" can be done for every Bot to ensure quick action for any incident reported. This will greatly assist anyone who is assigned to support at the time of the incident. Steps may be: contact the business; stop; rerun (with information about how much time to wait before rerunning: immediately, after X hours, the next day, etc.); let it run and update the business regarding the issue; etc. This "next step" document is shared with the COE, business, and technical team. These are the immediate first steps.

It should be very clear to everyone on the team that the person on the call isn't responsible for resolving the issue. That person is the contact person who will coordinate the response with the business, COE, and technical team. If the person on call is not able to resolve the incident, the person who is expert on that particular automation can be engaged.

Note   In RPA, the technical team is comprised of several different roles (process analyst, architect, business analyst, infrastructure analyst, developers, asset management, security, etc.).

If your organization has a COE, Level 1 support is maintained by COE, and Level 2 support is taken care of by the technical team. In the absence of a COE, both Level 1 and Level 2 support are the responsibility of the technical team. Once the Bot is deployed, Level 1 support starts monitoring the Bot.

3.2 Monitoring the Dashboard

After Bot deployment, it must be maintained to assure it is doing the work it was created for and that its performance is at an optimal level. Continuous monitoring is vital. Monitoring can be automated or manual. Automated monitoring could include sending alerts, such as emails or text messages, that advise if the Bot is underperforming or has stopped working or by updating dashboards.

Dashboards are very handy for monitoring Bots. Most of the automation tools come with simple dashboards that can easily be configured for deployed Bots. Dashboards are easily accessible via user interface by the technical teams or business partner to keep an eye on the real-time status of the Bot and how it's performing. As Sharda Cherwoo and Roy Rachamimov have said, "The operational dashboard is vital for the day-to-day running of the bots - monitoring their activities frequently and in real time. At the early stages in an RPA's program, there are few bots and all the attention is on integrating them into workflow and making sure they complete their tasks successfully."[1] (Note: for more information, please see the article referenced in the footnote.)

Prioritization

This is when many Bots need to be rerun. For example, servers on which the Bots are running are down, and as a result, there are many Bots to be rerun.

---

[1] www.linkedin.com/pulse/what-gets-measured-done-creating-rpa-dashboards-achieve-cherwoo/?articleId=6628417715835060225.

Prioritization will come from the business owners, determining which will come first. Some Bots are processing business-critical work (e.g., satisfying regulatory requirements).

In the case where there is no business-critical Bot to be rerun and all the Bots have the same priority, it is the decision of the COE or tech team. They can decide to run first those that are very short or have a small amount of work to get those completed. More complex Bots that will take more time to run will then be run. The guidelines for prioritization should be established up front, but should be sufficiently flexible to adapt to differing or varying situations.

Warranty

Whenever a new automation is deployed to production or a new release is moved to the production environment, there is a warranty period assigned to the Bot.

The warranty period could be of one week or two weeks as per agreement between the business and technical teams. Bots take time to stabilize and there might be some issues initially with new Bots. So the warranty period allows for the solving of defects quickly fixed with minimum dependency on multiple people. During the warranty period, all the documentation required to move the Bot to production, such as business sign-off, testing sign-off, etc., is not required. The fixes are done quickly to keep the Bot running. As mentioned earlier, the warranty period can be extended if either the business or technical team feels that the Bot needs to be closely watched for a longer period of time.

Bot Categorization

Bots can be "critical" or "noncritical." Critical Bots are those which are time critical; their output may be required for government regulations. These Bots have a significant impact on the business. If a critical Bot fails, the technical team must be able to take action immediately, without waiting for a decision from someone else to do so. For critical Bots, all the actions to be taken in case of failure must be defined prior to deployment.

Noncritical Bots do not have the same impact on the business. In case of failure of a noncritical Bot, advice from the business is required before any action is taken. The business will decide whether to rerun the Bot or take some other action. The technical team will respond accordingly.

Communication

In Chapter 5, we introduced the need to be in communication with your company's COE (Center of Excellence), if there is one, or, if not, with the department that will be responsible for deployment. Now that you are about to deploy, the relationships you established earlier by involving this group and letting the people there know what was being deployed, when deployment was desired, and other pertinent details, will greatly assist you in getting the Bot up and running when required.

In earlier phases, you advised this group (or individual) of the process being automated, what systems it would interact with, how often it would run, what platform it would run, and other pertinent details. They knew what date you wanted deployment and have, hopefully, worked to assure that everything was in place on their side. You need now to update them on a variety of issues, since things can change from day to day. Here are some of key topics to confirm:

- Business unit requesting the automation
- Requested date
- Impacted systems
- Date of Governance Committee meetings
- Any changes that the business might ask for after the additional request
- Anticipated size of the Bot (number of steps being automated, volume)
- Frequency of run
- Others as may be required by your team and the particular Bot being created

Change Management

When Bots run, there are always issues, and change management is used to fix, deploy, and keep track of them. Tracking of changes requires the creation of change tickets so that the work can be documented and tracked. There are many situations where different types of actions to fix a problem are required. As a result, change requests can be categorized to facilitate the team in taking the correct actions.

Tickets can be categorized as a change request, a service request, an investigation request, or a general request.

A general request is created when information regarding any Bot is required, such as the total work done, when the Bot started or completed, etc. For these types of requests, there is no need for any approvals since they are informational requests only. There is nothing immediately to fix change. These requests are triggered by the receipt of any alert produced by the Bot. The information is provided to the technical team for review.

Following the review of the information provided by the general request, either an investigation request or a change request may be created. An investigation request is created if additional resources are required to further explore the issue and find its root cause. In most cases, the Bot performed fine in the test environment, but isn't performing as well in the production requirement.

Some general requests may trigger a change request. The technical team may receive a general request as noted earlier. If they are able to identify the issue and the root cause, and they can implement the fix, a change request will be created. Following approval of the change (generally done through email), they can update the Bot code files.

Pause and Consider   These are the types of requests that are generally used, but it depends upon your organization's needs. How do you think you should categorize tickets in your new RPA program? If you are unsure, use what we have here; your program will evolve and you will learn over time if these categories best suit your organization's needs.

Tip   Having too many categories of tickets makes it difficult to understand and manage expectations, and it can become difficult to discern one type of ticket from another.

Auditing Bots

Once the Bot is deployed and is doing its work, we start collecting information about Bot performance. There is much information to glean.

- How many work items a work is processing in X amount of time? The expectation will have been determined earlier in the process.

- What is the workload the Bot is receiving every day? When the request was initially made, an estimate of the workload was provided. This probably changed as you learned more about the process, and now you will see what workload the Bot is actually receiving.

- How many times does it error out during the day? This indicates that an error has occurred; the Bot may continue or stop, depending on the error. Sometimes the Bot will encounter an error that it can move beyond, but sometimes the error will cause the Bot to stop. Depending on the error, the Bot may try to process it again.

- On-time starts vs. delayed start. This includes how many times the Bot starts on time, and if the start is delayed, how long was the delay. It also includes whether or not the Bot completed the work if the start was delayed. Knowing this information may lead to rescheduling the Bot start time.

- For complex work items with many steps, information is collected on sub-items between the works. This means looking at the individual steps at the smallest level and determining where the issues causing problems exist.

- Once this information is gathered, it can be analyzed and verified where the Bot is taking more time than expected; if erroring out too frequently, or infrastructure is slow, or applications are slow, these can be analyzed and addressed.

As noted in Chapter 4, an initiative could be a new Bot, an enhancement to an existing Bot, or defect correction. Deployment to production is slightly different for each case. Most of the information in this chapter refers to deploying a new Bot.

For an enhancement, a lot of the required information will already exist. You will generally know the systems and applications that the Bot accesses, and an enhancement will probably use the same ones. If there are changes, they will be noted. An enhancement often includes automating additional steps within the process; for example, if a Bot automated steps 3–9 of a process with 15 steps, perhaps additional steps are requested to be automated. The process is already known, so that need not be documented again. So the deployment steps are basically the same, but are expedited because so much information is already known; you are just building on what already exists.

This doesn't mean, however, that all identified risks need not be mitigated or that business doesn't need to sign off on the enhancement. But many things for deployment will already be in place from the original deployment of the Bot.

Defect fixes can be deployed by simply assuring that so doing won't "break" anything that is working. Testing sign-off is required, and then the fix can be put in production.

■ **Tip** If the defect occurs outside of warranty, testing sign-off and approval of updated requirements may be required. This depends on the standards set in your organization, but both approvals are recommended.

Disaster Recovery

This is the plan required should the Bot fail to do the necessary work. This could occur because servers are down, the Bot is working too slowly, or applications that the Bot is using are unavailable (e.g., Bot is using Jira, but Jira is unavailable). In these and other related cases, the functions of the business must continue. The disaster recovery plan, detailed in the SDD (see Chapter 6), is the blueprint for use in these emergency situations.

If there is a COE, that team will notify the technical team that a Bot is not working. In the absence of a COE, the business or the technical team itself will identify the problem due to its monitoring of the Bot, or the business may notify the technical team. The technical team and the business will decide to either run the process manually or wait until the issue that the Bot has encountered has been resolved and then rerun the Bot.

Retirement of Bots

One of the things that must be planned for is the eventual termination of the Bot. There are many reasons why a Bot may no longer be required.

Sometimes, the process is terminated. For a variety of reasons, the process may eventually no longer be needed. Perhaps through a merger, the purchasing company's process has incorporated the work of the Bot. Or the Bot may have supported a service that the company no longer offers. There are many reasons for a Bot to be determined to be no longer necessary.

In some cases, the current process has changed so significantly that a new Bot is required. All organizations are constantly striving for improvement, and processes will need to evolve along with them. Adjustments to the Bot can certainly be made, but there may come a time when it will be more economical to terminate a Bot and replace it with a new one. This becomes a simple "dollars and cents" decision.

There are situations where the automation software is converted to another software tool. Sometimes it may be necessary to convert from one software tool to another. This could be because the existing tool is unable to meet the automation need, a less expensive tool has been located, or other reasons. The Bot will be "retired" for the existing tool and a new Bot built for the new tool. The new Bot may be the same as the retired Bot.

Despite all your careful planning, sometimes the Bot is simply not achieving minimum savings. During auditing, you will be able to see how effective the Bot is. Despite all your due diligence through the stages as detailed in the previous chapters, it's always possible that the Bot simply doesn't deliver what it was meant to. You may want to make some adjustments, but, even by so doing, you and/or the business may come to recognize that the Bot simply isn't worth its expense. If your organization has a COE, someone from there will advise you that they have reservations about the Bot's performance.

In this situation, you will need to work with the business (and the COE if there is one), to determine if some adjustments could be made that would improve Bot performance, or if simply terminating the Bot is the most cost-effective decision (please bear in mind that cost is not always the determining factor of keeping or terminating a Bot. If, e.g., it is providing a service that enhances the customers' perception of the company, it may be advantageous to maintain it, even if doing so is an expense).

Steps to retire the Bot

Retiring a Bot is something that will be done eventually for most Bots. Initially, this will probably not be an issue, but as conditions and technology change, there will be times when a Bot will no longer be needed.

The following are the steps leading to and actually retiring a Bot.

- A request is received from the business to retire a Bot. The retirement of the Bot is the decision of the business. Information may have come from the COE that the Bot is not performing as expected, but the decision to retire it rests with the business.

- Identify the stakeholders. This is required since many areas are probably dependent on the Bot; either they are responsible for providing information to it, or they receive information from it.

- Collaborate by clearly stating which Bot is to be retired and which processes are impacted. It is vital that there is no question about the Bot being retired.

- Determine when execution of the Bot will stop. This will be done in conjunction with the business and other pertinent stakeholders.

- Identify any user IDs or access that must be terminated. Different roles may have been able to access various systems when the Bot was in operation, and now that access must be terminated. Please note that there may be cases where some or all of the access will not be terminated. Work with the business to know these situations. Be aware that if the Bot is being discontinued and the process it had automated will once again be handled manually, access to some or all of the systems may need to be available to individuals.

- Determine alternative options for when it will stop doing the work. If the process is being discarded, no options are required. But if the process is going to continue but be performed manually, the business must have sufficient time to make the necessary arrangements and operationalize the manual process. This will probably have no impact on the RPA team, except in determining when the Bot can actually be retired.

- Obtain the necessary sign-offs for retiring. These would include business representatives and the COE (if there is one).

- Collaborate with COE (if there is one) to actually stop the Bot and release all user IDs and resources. If there is no COE in your organization, the technical team is responsible for stopping the Bot.

- Officially declare retirement of the Bot. Send an email to all pertinent stakeholders, advising them that the Bot is no longer operational.

■ **Pause and Consider** As shown, there are several reasons for Bot termination. Can you think of processes in your organization that may not be particularly profitable from a balance sheet point of view, but are still important? Be sure management is aware of any such processes that are being automated, so they understand that there may not be a huge financial advantage in the automation, but that it is important nonetheless.

Case Study

Now we will look at how these tasks would be performed in the hypothetical case study we have been following.

Deployment to Production

Item	Item Details	Comments	Status
Product (Bot) Name	Phone and address update	This name will be used in inventory, reports or change requests, etc.	Complete
Deployment date and time if applicable, agreed upon	Jan 1, 2022	This has been agreed upon with all necessary stakeholders.	Complete
Size of automation	Medium	Based on volume and frequency of run, we have determined the size as medium.	
Bot schedule	Monday to Friday: 3 times of day 10 am, 1 pm, 6 pm	This satisfies the needs of the business as agreed upon by the product owner.	
Contacts during warranty	John Smith Mary Jones	The developers named will be available during the warranty period to address any issues that may arise.	
Statement of segregation of duties	Code written by Mary Jones Documentation Sarah Williams SME involved in UAT: Douglas Johnson and Betty Parker	The individuals named will be available as needed during the warranty period.	
Volume of work	100 cases for one run		
Access verification	Access verified for web application XYZ and for mainframe ABC	The developer named earlier has been granted full access to both the XYZ application and the ABC mainframe. This will minimize delays if problems arise with the Bot.	
Email addresses for communication	Dev1@company.com sem1@company.com	These emails will be used to provide the stakeholders with a daily update of Bot performance and to advise them of any issues. There will be once-a-day emails if there are no issues, but if issues arise, these stakeholders will be notified immediately at these email addresses.	

Item	Item Details	Comments	Status
Disaster recovery	Stop the Bot and inform contacts.	The business will decide whether or not to perform the process manually while the required repairs are made to the Bot, or whether to rerun the Bot after the issues are resolved.	
Major/minor version	1.0.0		
People who will be present at deployment	John Smith		
COE contact		There is no COE.	Not applicable
Folder structures	No new folder need to be created		Not applicable
System/application accessibility	Application XYZ, mainframe ABC	Both are accessible by the required parties.	
Infrastructure capability	Phone and address update Bot is available to execute.		Complete
Risk identification and roles responsible	All identified risks have been mitigated.	The business owner assumes all responsibility for any unanticipated issues that arise.	

Deployment instructions/steps
 	
Task Deploy Steps
  – **Pre-implementation steps**
  – **Implementation steps**
  – **Test the implementation steps**
  – **Backout steps**
  – **Post-implementation steps**
		Roger Adams, the Automation Services Deployment Manager, has specified the implementation and backup steps. These are documented in the form "Address and Phone Update Bot, December 1, 20xx," available at "C/ Business_Unit_123/Bot_ Directory/Address_and_ Phone_Update_Bot.doc."	

(continued)

Item	Item Details	Comments	Status
Production Support Model	Both Level 1 and Level 2 support are the responsibility of the technical team	Automation services will engage the technical team as part of the deploy-to-production process to ensure a smooth transition to maintenance. This will include a thorough knowledge transfer.	

■ **Pause and Consider**   The checklist is extensive. It may be complete for your organization; it may have items you need not consider, or it may not have factors that will be important to you. Based on what you have learned about RPA, and your knowledge of your own organization, what do you think you might need to add to the checklist? What do you think can be omitted? If you are unsure, use it as is for your first few projects. Time will teach you what you need to change.

Now that you have deployed your Bot, we will look at organizational structure models. You will have used whatever the model currently is within your organization, but as your RPA program matures, you may want to make changes. Chapter 8 details the most common organizational structures.

Organizational Structure

One very important aspect to remember about Robotic Process Automation software is that it doesn't require companies to have a full organizational makeover nor a huge IT transformation. Simply stated, there is no need to make any radical changes to the core business processes or existing back-office technologies. For attended Bots, standard desktop application software is installed, and for unattended robotics, a virtual desktop server can be set up, or RPA can be integrated into an existing virtual desktop environment. All this makes it very easy to quickly implement RPA in any organization, despite its size or structure.

When rolling out an RPA program, rather than focusing on tool selection, focus on team enablement (team structure, training, etc.), the operating environment, and on overall RPA governance. Each organization will have different needs, and you need to assure that your new RPA program fits your organization. Selecting RPA software will be easier after you have the RPA structure established.

■ **Pause and Consider** How will you structure your RPA team? What roles will be required in your organization? What training will each person require? Who will decide on an RPA tool? How will you recommend a tool? What research will you do? Giving these issues careful consideration will assist you in successfully implementing an RPA program in your organization.

R. Fantina et al., *Introducing Robotic Process Automation to Your Organization*,
https://doi.org/10.1007/978-1-4842-7416-3_8

The RPA program's organizational structure provides the model of the framework for deploying RPA at the enterprise level. As we addressed in Chapter 3, choosing the right path for an RPA program will make a huge difference between a successful RPA program and failure that leads to inadequate results and ultimately high cost. These issues could cause withdrawal of executive sponsorship and ultimately cut off financial investment in RPA. The organizational structure of RPA-related functions in an organization is important, because it is not only related to costs but also to decision-making that will define a long-term vision and drive strategy to execution. Who makes those strategic decisions is clearly an important factor in how well an organization achieves its business value.

The optimal organizational structure will depend on an individual RPA program strategy, size of the serving population or business operations, operations complexity, management culture, organizational design, risk tolerance, and many other factors.

In this chapter, we will review in depth the three most common structures that were briefly introduced in Chapter 3: centralized, decentralized, and hub and spoke. We will fully describe their benefits and the possible drawbacks of each. We also will discuss which RPA program functions could be outsourced and which should be in-house, required architecture and support for the robotic operating environment, and how they all fit into the enterprise IT architecture.

We will also help you to understand how to determine the right degree of centralization and decentralization for your organization and how the organizational structure may change as the RPA program becomes more mature.

For any RPA program, the following functions are required:

- *Architecture of the robotic operating environment* (infrastructure support, technology choice)

- *RPA operations* (maintenance, support, monitoring, training of new resources, change management)

- *Governance and strategy* (integration to overall organizational structure, oversight of resources responsible to support all functionalities of the RPA program, compliance to policies and procedures, security system access, process prioritization, escalation path)

- *Delivery* (process discovery and assessment, solution design, testing, deployment, development of standards)

Centralized

In a centralized model, the functions mentioned earlier are part of a centralized IT division or unit that is under single executive leadership with full control on decision-making, defining strategy, and with sufficient financial resources.

This IT division is typically designed as a Center of Excellence (COE) that is established within one executive office and manages the entire life cycle of the RPA program: defining strategy for RPA on the enterprise level, technology choice, architectural and operating infrastructure, operations support, delivery RPA solutions, developing standards, and ensuring that adequate controls, risk management, and compliance are in place. This means it oversees all functions required throughout the end-to-end life cycle of the RPA program.

An effective RPA COE means far more than establishing a generic IT team consisting of skilled and experienced developers. It requires the creation of a program consisting of many different roles that need to be filled with the right people who will fulfill all critical tasks required for each function. Those roles include the following: RPA program sponsor, operations and program lead, project manager, RPA/business analysts, solution architect, developers, IT infrastructure engineers, operations support and service staff, and risk and control team.

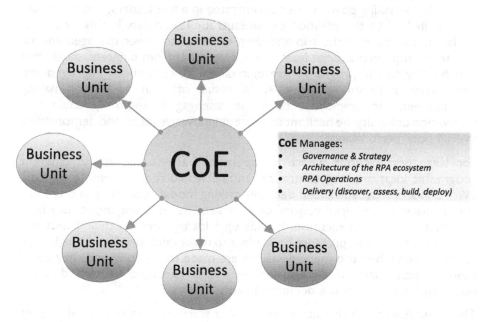

Figure 8-1. Centralized model

Benefits

A centralized model works best for companies that need greater control in terms of the RPA strategy, cost, execution, and visibility. This structure leads to lower costs associated with hardware licensing by having all that is needed in one place. When it is distributed across the enterprise, oftentimes extra or duplicate equipment and licensing are needed that would lead to increased redundancy and inconsistency in methodology and tools. This can contribute to the technical department budget and ultimately increase overall costs of the program.

The centralized model gathers under one umbrella the collective resources and expertise that are required to deliver the RPA implementation successfully. This enables those in charge to view all initiatives in a centralized place and gives them stronger governance abilities over projects and priorities. It enables an end-to-end view of process changes, leading to more beneficial opportunity identification. A central model also provides a standard set of regulations and practices for assessment, delivery, monitoring, and maintenance.

Ultimately, all of the above attributes make scaling the program easier.

Disadvantages

As with any type of centralization, this model establishes a structure in which the decision-making powers are concentrated in a few leaders, and this could potentially lead to the creation of bureaucratic leadership. In this structure, decisions are made at the top and then cascaded to lower management for execution. It prevents, or at least limits, employees from different areas from contributing to the program's strategic direction, continuous improvement, and overall program maturity. As a result of such a decision-making environment, the employees could be disengaged, and they could lack motivation or loyalty, be hesitant to present innovative ideas, and demonstrate overall poor performance.

For large organizations where its operations are spread across different geographic locations, the centralized model would require remote control. With a lack of a decentralized decision-making model, leadership is often not able to have the required control over implementation, nor might they have the time to manage execution. It adds up a lot of work on their hands and creates a tremendous pressure on leaders to make decisions fast. It might lead to a situation when too many decisions are made too quickly, and they could then be inadequate or disregarded or poorly executed by staff that are remotely located from the decision-makers.

The centralized model is suitable for small or startup organizations with fewer business units or one business division. It brings speed, scalability, and cost-effectiveness. It also could be used as a start-up point or proof of concept (POC) for a large enterprise where the organization wants to pilot, learn, and validate the value of RPA. It would help executives to make decisions on keeping and expanding this capability from POC to full-scale implementation across the entire enterprise. However, you might find that a different model, decentralized or hub-spoke, would suit your needs better.

Regardless of the appealing advantages of the centralized model, before you proceed with it, be mindful of its disadvantages since they might stall the adoption of your new RPA program and result in a poor ROI.

■ **Pause and Consider**   Based on the information you have just read, what do you think of the centralized model for your organization? Does it resonate with you? Or does it sound foreign? There are two additional models to consider, but give some thought to this one.

Decentralized

In a decentralized model, multiple individual RPA programs are established within an enterprise, with individual RPA frameworks operating under different executive branches or business units. This can be viewed as multiple sub-COE offices that are operating autonomously. The decentralized model has its functionalities spread across an organization with different capabilities being run by different business units. In this model, decision-making and operational management responsibilities are owned by the local RPA teams that are overseeing all required functions, from defining the architecture, the operating environment and governance, to delivery and maintenance.

This model places fewer constraints on local business teams within the organization, while simultaneously helping them gain momentum and expertise. It hands the demand for innovation over to the employees by empowering them to meet business goals by using RPA. This model is loosely governed, and different lines of business establish their own guidelines, standards, methodologies, and structures.

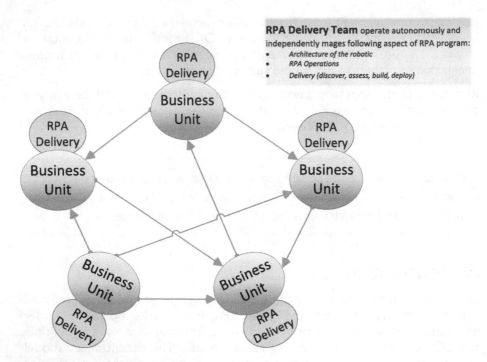

RPA Delivery Team operate autonomously and independently mages following aspect of RPA program:
- *Architecture of the robotic*
- *RPA Operations*
- *Delivery (discover, assess, build, deploy)*

Figure 8-2. Decentralized model

Benefits

Innovation thrives in those conditions where decisions are made by lower levels and smaller groups of decision-makers. Greater autonomy empowers people; it gives them a sense of importance and ownership, since they feel that they are accountable for steering the direction of the organization. Employees become more loyal because they are allowed to take personal initiatives in day-to-day work with minimum managerial approvals. There is nothing more satisfying than seeing the successful results of your own decisions. Most importantly, it also allows them to leverage their knowledge and expertise and be able to experiment and learn: key components in driving innovation forward. All those features create a self-sufficient and self-maintained structure, where employees at all levels are accustomed to work autonomously, take a lead, and make decisions when needed, without having to wait for a decision to be made further up the chain of command.

More efficient decision-making brings speed and agility to the organization, to assist it in succeeding in today's fast-paced and ever-changing business environment.

Disadvantages

While the decentralized model is a great way to empower employees and drive innovation and business agility, it could potentially cost more. Additionally, it may be difficult to liaise with enterprise-wide IT personnel, since there is no central control to define strategic direction.

Eventually it could lead to misalignment and a major disconnect from the overall enterprise strategy. Multiple RPA teams with their own authority to define their own strategy might lead to different architectural structures and different vendors. This could ultimately lead to having, within the same enterprise, different technological capability, tools, operating models, methodology, and standards. It also could create redundancy, since the same functions could be duplicated across the enterprise, leading to higher overall RPA program cost.

In addition, it might create silos or islands, where multiple teams are solving the same problems. To keep those islands connected would require strong relationships and efforts among leaders to create some sort of "community of practice," where best practices could be shared. However, employees involved in supporting RPA teams could be so busy with their day- to-day duties that they easily miss the opportunity to take advantage of those shared practices. In the eyes of an employee, when under pressure to deliver, it is easier to reinvent the wheel than spend time to research, understand, and reuse best practices. It sometimes seems preferable to create something new simply because existing solutions may not be known or are not quite suitable or easily transferable due to differences in tooling, development practices, or even the nature of the business operations.

Because of all those features, generally the decentralized model is more expensive to run, less efficient on the enterprise level, and is apt to create technical debt which sooner or later needs to be addressed.

For a small business, rapid growth may create the need to decentralize to continue efficient operations and maintain business agility. Despite its advantages, the delegation of control to other decision-makers may be difficult for business owners who are accustomed to making all the decisions by themselves.

The decentralized model is suitable for large enterprises with multiple divisions or multiple business branches, where the nature of their business operations are vastly different and self-maintained. Usually, we might find that technology services are already decentralized and are aligned under their own business branch or division. In those cases, most likely, they have different core architecture, operating and delivery models, systems, and tooling with their strategy aligned to meet the specific business needs they are part of. Therefore, RPA and business leaders need to ensure that they prioritize which processes will initially use RPA and develop a transformation roadmap based on the individual requirements.

■ **Pause and Consider**   You have now been introduced to two structural models. Which one seems to fit your organization better? Does one stand out, or does it seem that neither would be optimum for you?

Hub and Spoke

The hub and spoke is a model that resembles the spherical shape of the bicycle wheel, where the hub is in the center and spokes are connected and meet at the hub, which is the center of the wheel. This concept was pioneered by the transportation industry, and today, it is widely adopted by different companies in every industry.

In this model, the COE is established within one executive office and typically serves as a hub for decentralized spokes. The delivery team can be seen as spokes.

The COE sets standards, develops policy, provides training, and focuses on innovation and RPA program strategy. In this case, you can view the Center of Excellence as the main source or library of knowledge and expertise, including best practices, frameworks, development practices, reusable components, training, and as an IT unit that supports, maintains, and evolves RPA technical infrastructure and capability.

Spokes – RPA delivery teams – are primarily focused on identifying opportunities, developing, and deploying RPA solutions to the business area they are assigned to or support. This model requires close collaboration between the hub (COE) and spokes (RPA delivery teams).

In this model, the COE would

- Ensure integration of the RPA operating environment into the overall corporate IT ecosystem by creating a collaborative model with IT system/application support and solution architecture teams

- Define the RPA program evolution roadmap and drive strategy to execution

- Drive innovation and introduce new technical capabilities

- Maintain and support RPA-related IT infrastructure

- Develop standards, automation blocks, or reusable components

- Interact with the RPA vendor and service provider

- Be accountable for technology and tooling choice
- Manage and maintain a library of reusable components and best practices
- Train and enable delivery team personnel
- Ensure knowledge is shared and reused
- Report on RPA program performance and effectiveness on an executive level

Delivery teams would

- Assess and prioritize selected processes to be automated
- Develop RPA automation and place Bots into production
- Interact with IT teams such as system/application architecture and support
- Provide change management support to business units
- Ensure existing robotic workforce performance
- Maintain existing automated processes
- Perform security/compliance functions

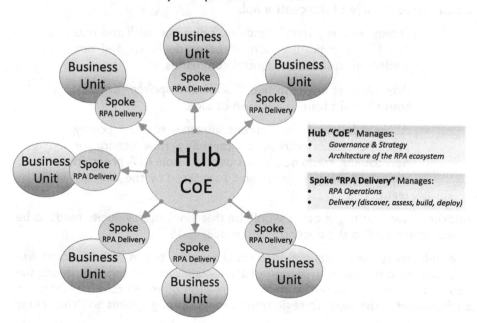

Figure 8-3. Hub-and-spoke model

Benefits

This model takes some of the burden of daily business operations off the COE. This includes allowing other teams outside of their reporting structure to perform such tasks as identifying processes suitable for RPA and developing and monitoring the performance of their own Bots. This frees COE resources to spend more time on big-picture items, such as planning for program evolution, introducing new capabilities, working on tools, knowledge sharing, and recycling activities such as the development of best practices, reusable components, etc.

The model is cheaper than the decentralized one because due to centralization of certain functions and activities, duplication can be avoided throughout the entire organization. For instance, instead of having four different people across the organization looking after RPA-related IT infrastructure, here, it might be sufficient to have two.

Disadvantages

The most common challenge with the hub-and-spoke model is that it relies on perfection, coordination, and collaboration. When a whole model, hub in the center and every spoke, is working in harmony together, then the entire RPA program is going to be incredibly efficient and effective. If just one spoke were to get out of place, then the entire model could start to experience problems. Eventually, more spokes are going to fall off and the whole RPA program becomes ineffective and costly. There can be different causes that create issues outside of the central hub:

- Spoke "delivery team" leads may disagree with the hub "COE" strategic direction or other decisions and not implement or execute central commands.

- Any mistrust between hub "COE" and spokes "delivery teams" might lead to creation of silos.

- A breakdown of the current infrastructure or poorly implemented or introduced capability, new vendor or new tooling, could be disastrous to a whole RPA program impacting every business unit supported by the delivery team.

Another issue that might occur would be that innovation velocity tends to be slower compared to the decentralized model.

The hub-and-spoke model provides many benefits, but in order to capitalize fully, proper functional distribution of accountabilities and services among the hub and spokes is required. In addition, creation of an effective collaborative environment, trust, and strategic transparency among spokes and the center hub "COE" is essential to the success. The model isn't for every organization, but many might find some of its components useful by tailoring their mandate and roles to the hub (COE)-and-spoke (delivery team) model.

■ **Pause and Consider**   You have now been introduced to the hub-and-spoke model. Do you see it as successful within your organizations? What components of it seem workable, and which do not?

Each of these models can work successfully at any organization if applied appropriately. There is no definite winner; no single model can be seen as the most or least effective, as it depends entirely upon the type of the organization, nature of their business, service or products, culture, size, and reporting structure.

■ **Pause and Consider**   Based on your knowledge of your organization, and considering how receptive you think both management and line workers will be to the introduction of an RPA program, what do you think the organizational structure in your organization should be?

What to Consider Before Selecting an Organizational Structure

As we described earlier in Chapter 3, before setting on a RPA path, the program leadership should decide what functions that are required to support and execute the RPA program are going to be led by in-house resources or be outsourced to an external third-party service provider.

The more companies are experienced with the usage of RPA technology, the more they tend to use it in-house. They have accumulated sufficient knowledge and technical expertise so that it might be more effective and efficient to keep those resources within company. On the other hand, those companies that are new to RPA capabilities are more likely to outsource to their RPA external partners.

This is a very interesting phenomenon, and both options offer different benefits to each party.

A 2016 study conducted by Capgemini Consulting and Capgemini Business Services "indicated that while companies will gain benefits implementing RPA with help of an outsourcing partner, the benefits could not be as high in comparison to companies that make the technological investment in RPA themselves."[1]

However, starting an RPA program from scratch would have higher costs since it would take time and investment to reach adequate maturity and realize financial benefits.

Nevertheless, for companies that are not certain and not ready to commit to making the technological investment in RPA themselves, but still want to leverage RPA technology, they would benefit in partnering with an external provider. They need to keep in mind the required RPA functions that we discussed earlier in this book while selecting their RPA partner. It is important to select and partner with a provider that could provide a full-stack service, from process discovery to Bot deployment and maintenance. Companies that choose to follow that model do not need to make significant investment in technology infrastructure or be concerned about identifying the right processes and the sufficient number of processes to gain an adequate return on investment. Now those concerns are outsourced, and the company can terminate the partnership at any time if it finds it is not suitable or does not bring enough benefits.

However, as was found in the Capgemini study, "it is clear that those who implement RPA in-house will benefit the most in the long-term."[2]

This is why it is vitally important to define the organization's strategic course and long-term vision prior to embarking on an RPA journey.

Also, an organization can choose a hybrid model that establishes a collaborative framework with an external provider and in-house resources. For instance, an external partner could play the role of a delivery team, focusing on suitable-for-RPA process identification and assessment, development, and deployment of the automation solutions (Bots). The in-house team would then provide support and maintenance.

The following tips could be helpful for choosing a suitable path.

Prior to deciding which model to follow, an organization must be clear on these aspects:

- Are there enough processes for RPA to justify investments (ROI)?

- Is the organization committed to long-term investment in an RPA program?

[1] https://drive.google.com/file/d/1H2YLKUDkkCq6PFsECgCe1W79prYCRyBM/view?usp=sharing.
[2] Ibid.

- Does the organization have the in-house skills and capability to start an RPA program?

- Is RPA a part of the long-term (meaning five- to ten-year) strategy of digital transformation of its operations and IT infrastructure or ecosystem?

Organizations must ensure that they consider all the aspects of RPA program implementation in terms of operational culture, people, processes, and in-house technology prior to selecting the suitable model for RPA program organizational structure.

Keep in mind that the RPA operating model is one that will continually evolve as the organization matures with time. Therefore, companies don't have to pick one model and stick to it. It might be even beneficial to start small, perhaps as proof of concept with a centralized team structure, to support one business unit to learn, explore, and evolve organically to bring more benefits and slowly grow and scale up across the enterprise. Another factor that could trigger the need to change is the fast pace of technology change. Automation capabilities evolve rapidly, maturing an organization into more advanced stages of automation. That level of maturity could require an entirely different operating model for the organization to maximize and leverage the full potential of that technological capabilities. We will discuss this in detail in Chapter 10.

Development Methodologies and Framework

Once the initial foundation for an RPA initiative has been established, it is time to begin work on the first project or series of projects. How that is actually done will depend on many factors in your organization.

RPA fits comfortably with traditional Waterfall methodology and Agile and its numerous hybrids and variations. How you proceed will depend on whether or not yours is an Agile shop or Waterfall.

Regardless of the methodology used, there will be some commonalities:

- The business case that was created earlier must be updated. Remember, the preliminary business case was based on limited information. By the time you are ready for development, additional information will have been obtained. This may change the specifics of what's required, the anticipated savings, the other systems impacted by the proposed change, the tools used, etc. These changes must be incorporated into the updated business case.

© Robert Fantina, Andriy Storozhuk, Kamal Goyal 2022
R. Fantina et al., *Introducing Robotic Process Automation to Your Organization*,
https://doi.org/10.1007/978-1-4842-7416-3_9

■ **Pause and Consider** How is this information best gathered? What meetings have been held up to this point that have included information that can be used to refine and update the business case?

- A readiness checklist will be required. This is used to assure that all the steps necessary during development to deploy the software to production have been completed. The readiness checklist follows the project through its life cycle, since there are requirements for deployment that must be accomplished at each stage.

- Roles: When work begins on an RPA project, several roles will be required.

  - Business analyst: This person has been involved from the start, working with the business to fully understand the need and translating that need to requirements.

  - System analyst: This person has also had some involvement from the start, working with the business analyst to identify the systems impacted by the proposed change. Early in the project life cycle, exactly what systems are impacted may be suspected but not confirmed. During this phase, the exact systems impacted must be identified.

  - Architect: The architect is also not new to the project, having worked with the business and system analysts to help determine a general timeline for the project and to evaluate the level of complexity. Like the business and system analysts, the architect's work now becomes more specific, as the "unknowns" from early in the project become clear.

  - Developer(s): A developer may have had some input early on, but as long as there was an architect engaged, it is unlikely. Now the developers will read and understand the requirements and translate them to code.

  - Business owner: The business owner is the requirements expert. They know what is needed and assist in creating the iterations of the project.

If Agile (or one of its hybrids) is used, the previous roles will all be involved, but simultaneously rather than consecutively. RPA lends itself very well to Agile, due to the iterative nature of both.

Challenge with Differing Methodologies

RPA teams usually work with more than one business unit at a time, because they are working with different applications which may be owned by different business units. The RPA team cannot change its own methodology for every specific client or utilize multiple methodologies at one time. If there are different methodologies that other business units are working with, and RPA is using, for example, Agile, there are challenges: how can one methodology be used by RPA when the other business units are using a Waterfall methodology?

In the financial sector, for example, many business units are using a Waterfall methodology, while most RPA teams are using Agile or Lean. The business unit is accustomed to signing off on completed requirements; Agile and Lean use an iterative approach, wherein partial requirements are approved and then developed. The RPA team then has to wait until all requirements are approved to begin development. How can the wait time be minimized?

Instead of signing off the entire requirements document, the business must sign off on sections incrementally, noting in the document that the sign-off does not pertain to the entire requirements document, but only to the particular section. Upon full completion of the requirements (note that by this time, most of the development will have been completed), the business can then sign off on the entire document.

This requires training of members of the business unit. The concept of signing off only partial requirements will be new to them, and they may not be comfortable with the idea. They may say that they can't possibly sign off on "partial" requirements, because they need to see the whole picture before they can approve any part of it. This is, certainly, for some, a new way of doing things. For analysts and for SMEs, they need to think about how the requirements document is divided. They will need to look only at the steps that are being automated, but even then, they may want to see the end-to-end requirements document before approving any one part of it. This is true of deployment as well; the Bot will be deployed incrementally. Having a product RPA incremental development roadmap or plan is critical to define small pieces of RPA. These are building blocks for the final automation. This roadmap might include epics, stories, or other artifacts. Regardless of what they are called, they constitute the step-by-step roadmap for the automation.

■ **Pause and Consider**   Each portion of requirements should be large enough to develop or be a complete functional portion of RPA, for example, accessing the email account to retrieve email content, or accessing the system's/application front end by logging in under RPA credentials, etc. Each requirement should produce a "minimal viable product," even if that small "chunk" is never delivered separately to the customer. How can you best work with the business to help them understand this division or "chunking" of requirements?

Consider again our case study. A client sends an email notifying us that they have changed their phone number and address.

- First iteration of requirements: "Bot will open the email, read only the phone number, and update the phone number in the web application." The business will be asked to approve this first iteration of the requirements.

- Second iteration of requirements: "Bot will open the email, read the phone number and address, and update the phone number and the address in the web application." The business will be asked to approve this second iteration of the requirements.

- Third iteration of requirements: "Bot will open the email, read the phone number and address, and update the phone number and address to the web app and the mainframe."

In this situation, the first iteration of requirements, once approved, will be developed and tested. User acceptance testing will not occur until all components of the requirements are developed. The assumption here is that there will be incremental development, but only one release when all the products of all the requirements have been developed.

An option is that the product of the first requirements could be delivered and put into production, followed by the second being deployed, etc. This depends on the model that your organization selects.

With each new iteration of requirements, you are increasing the scope of automation. This may be a new concept for the business, which has been accustomed to only approving the entire requirements document.

In some organizations, the deployment team may be using a Waterfall methodology and the RPA team using an Agile methodology.

We will now discuss several common methodologies used with RPA.

Agile Methodology

Many RPA organizations select an Agile methodology or an Agile hybrid. The major aspect of Agile that is often incorporated into a new RPA team is the iterative nature of development, from start to finish.

In an Agile environment, some preliminary, very basic requirements are approved. This is the minimum viable product, meaning that the functionality works, but it may not be anything you'd want to give to a customer; it is too rudimentary. In our case study, a very preliminary requirement might be: "Bot will read the email, read the phone number, and update the phone number in the web application." The business would approve that, knowing that it would be pointless to give it to the customer, since it is only part of what the customer wanted and was promised. But it is still a basic function that needs to be done. Once that is approved, the development team would begin working on it. The next requirement may be: "Bot will read the email, read the address, and update the address in the web application and the mainframe." Again, this is only one aspect of what the customer requested and was promised, but absolutely required.

This, obviously, differs from the traditional, Waterfall methodology, in which all requirements are gathered and approved before any other work begins. But since, as has been shown in earlier chapters, RPA is best established following an iterative process (e.g., do some work; present to the Governance Committee; get approval to move ahead [or not]; do some more work; present again; etc.), using an Agile development methodology just makes sense.

One advantage is that changes can be made early in the project life cycle. For example, once the requirement "Bot must be able to read the incoming emails from mailbox XYZ" is approved, the business may decide that not only should address and phone number updates be done by the Bot, but name changes due to marriage, divorce, etc., should also be covered. In a traditional Waterfall methodology, this would be a major requirements change, requiring rework. In an Agile environment, since the requirements are captured and approved incrementally, such changes can be made on the fly.

■ **Pause and Consider**   Remember, the introduction of RPA in your organization can be revolutionary. If the methodology that the RPA team will use differs from what the rest of the organization is using, you will need to negotiate areas of difference. For example, how will you convince the business to sign off on partial requirements?

In comparing RPA initiatives using either an Agile or Waterfall methodology, Madan Kumar Divvela said this: "Typically, process automation with RPA takes a few weeks from inception to production. The traditional Waterfall methodology cannot keep up with the pace of RPA delivery."[1]

Benefits: One of the benefits of using an Agile methodology is that changes can be easily made without costly delays and rework. Commenting on his own experience, Dominic Whaley, writing in *Partnering Solutions*, said that "Dividing the work into distinct packets enabled us to take stock at regular intervals. Changes or new requirements were noted and added to the next sprint."[2]

Another benefit of delivering in increments is that lessons learned from the previous increment could be easily incorporated in the next set of requirements without needing to rewrite the whole set requirements and go through the reapproval process again. This way, something you think could have been done better in, for example, the gathering or documentation of the requirements for the first set of requirements can be done in the gathering or documentation of the second set.

This brings increased flexibility and reduces overall lead time of the development cycle.

Also, all the team members are busy all the time. Once the initial requirements have been documented, the developers and business analysts can begin working on the MVP. While this work is being done, the next set of requirements is being captured.

Agile also makes it easier to estimate the time it will take to develop each requirement, which makes scheduling deployment easier and more effective.

Another potential benefit was discussed in *Blueprint*, in an article entitled "How an Agile Approach to Automation Can Deliver RPA at Scale."

> *In an Agile approach to Robotic Process Automation, business processes are designed and optimized before any development begins. This allows large organizations to fully standardize and optimize end-to-end, complex, and multi-layered business processes where the real ROI in automation resides. It also provides them with the invaluable opportunity to consider and tie those processes to larger business objectives and enterprise or regulatory constraints, policies, and controls.*[3]

---

[1] www.linkedin.com/pulse/best-implementation-approach-rpa-agile-waterfall-madan-kumar-divvela/.
[2] https://k2partnering.com/robotic-process-automation/robotic-process-automation-implementation-waterfall-agile/.
[3] www.blueprintsys.com/blog/rpa/agile-approach-automation-delivers-rpa-scale.

Possible disadvantages: Working in an Agile environment requires training. There is not only a new way of working and an entirely new vocabulary but also a new way of thinking. Simply changing the vocabulary but using the same methods does not make an organization Agile. It is a characteristic of human nature that people don't like change, especially if they are very efficient and effective at the way they are currently working. Agile requires people on both the business and technical side to think about requirements, development, etc., very differently. While development teams often quickly get on board with Agile, the business side is frequently much more hesitant.

Pause and Consider   If your organization is not familiar with Agile, but you feel RPA would best be structured using this methodology, how would you work to convince people that it is? Do you see potential acceptance by the tech team, but resistance by the business? How do you get both groups to see the advantage of using Agile?

This is not meant to imply that only Agile is effective with RPA; that is not the case at all. While many organizations are moving away from the traditional Waterfall methodology, many are not. RPA can be established using this methodology as well.

Waterfall Methodology

Using a Waterfall (requirements, development, test, deployment) methodology with RPA simply replicates how projects are performed in any traditional environment. Once requirements are captured and approved, development begins. The process continues in a very traditional manner.

Let's look again at our case study, and consider it in the context of a Waterfall methodology within RPA. It would follow the traditional pattern: requirements gathered and approved, followed by development, testing, and deployment. The manual process is still being automated, so the end remains the same; only the means are different. Governance meetings would be held after each phase, to gain approval (or not) to proceed.

Benefits: Most organizations are familiar and comfortable with this very traditional methodology. There are no surprises; it is tried and true (if not the most efficient way of developing projects today). For some organizations, this can be the safest way of introducing RPA to an organization; the introduction of RPA will rock the corporate boat a little; you don't want to swamp it with an Agile tidal wave!

The Waterfall methodology works well for small and simple RPAs. If the process is very simple and straightforward, with little complexity, etc., the Waterfall methodology process will usually be successful. More complex manual processes are better automated in an Agile environment.

Possible disadvantages: This can be a much slower process. Work does not begin on development until all requirements are gathered and approved; testing does not begin until all development has been done. If, during testing, some error is found, or if the business decides it wants to add a requirement, much rework, with its inevitable accompanying delays, must be done.

It is only fair to note here that most of the literature comparing Agile and Waterfall usage in RPA programs comes down quite heavily on the side of using Agile. While that means you should certainly consider it, there are many factors you need to think about before making the decision for your organization.

Lean Software Development

Lean software development emerged as an alternative to the traditional Waterfall methodology with rigorous, document-driven, and traditional development approaches. Lean thinking principles are based on the Toyota Production System (TPS) and have been successfully applied in many manufacturing and product development organizations. Lean's philosophy is based on reducing the development time by removing all non-value-adding wastes.

As Taiichi Ohno, one of the main architects of the Toyota Production System (TPS), explained the concept: "All we are doing is looking at the timeline from the moment the customer gives us an order to the point when we collect the cash. And we are reducing the timeline by removing the non-value added wastes."[4]

It took Toyota 30 years to refine their methodology and practices to make them effective and efficient and what they are today. However, it did not get mainstream attention and recognition until MIT's (Massachusetts Institute of Technology) five-year study of the automotive industry identified Lean as a source of huge productivity differences between the United States and Japan.[5] The study resulted in defining a new term for production system – "Lean" (because by removing waste, it would require less of everything).

---

[4] Liker J. K, 2005, The Toyota Way.
[5] Womack J. P. et al., 1990, *The Machine That Changed the World: The Story of Lean Production*.

The Lean production methodology and thinking is mainly described in the book *The Machine That Changed the World* by James Womack, Daniel Jones, and Daniel Roos.

During the last few years, Lean has also become popular within the software industry. It mainly originated from the book *Lean Software Development: An Agile Toolkit for Software Development Managers* by Mary and Tom Poppendieck.[6] This book studies Lean production systems from a different perspective and adopts and tailors Lean principles and tools for software development. Its popularity is rapidly growing, and this growth is due to its effectiveness in identifying and eliminating waste and quickly responding to changing customer and market demands.

The key success of the Lean methodology is in the shifting of the practitioner's mindset to Lean thinking and not just adopting and practicing Lean tools. There are five core Lean concepts that underpin Lean thinking:

1. *Value* is defined by the customer and is what the customer is willing to pay for. It is critical to have a clear understanding of what that is. The product owner can be seen as the "customer" in RPA.

2. *Value stream* is a visual representation of every action (both value adding and non-value adding) currently required to bring a product or service through the flow from customer request to delivery. The process map, developed early in the life cycle, is a rudimentary but often adequate version of the value stream.

3. *Flow* represents how people and materials or required items are moving from one step to another until the value is delivered to the customer. It is important to achieve a continuous flow by removing waste from the value stream so development or services can move smoothly toward the customer. That would lead to gains in productivity and efficiency, making processes leaner. That would dramatically reduce time to the customer; instead of taking months to deliver, it could be reduced to weeks or even days. The document that you will create to accompany the process map will capture this information.

---

[6] Poppendieck M, Poppendieck T, 2003, *Lean Software Development: An Agile Toolkit for Software Development Managers.*

4. *Pull* means customers can order products or services when they need them, and those products or services could be delivered within a short time. It would ensure that nothing is built before it is needed, resulting in lower cost of inventory that potentially won't be used by the customer; all that effort of creating inventory would be wasted. You determine what the next item to work on is.

5. *Perfection* is probably the most important concept of Lean thinking. It requires making continuous process improvement as a core part of your organization's culture. Striving for perfection is done through the repeatable process in which any waste is continuously identified and removed. It is a long journey which requires persistence and constant effort on all levels, from employees, supporting teams, and leaders. As Mark Crawford mentioned: "Lean experts often say that a process is not truly lean until it has been through value-stream mapping at least half a dozen times."[7]

The primary guiding principle of Lean is the identification and elimination of waste from the process to maximize customer value. Note the concept of waste could be quite broad and would depend on the value delivery processes and environment. Let's draw parallels to the software development domain. For instance, waste could be defined as

- Extra features that the customer doesn't use

- Poorly designed functionality that failed testing and needs to be redone

- Waiting for requirements or support

- Partially completed work

- Task switching, working on multiple development initiatives at the same time

- Extra processing steps or lines of code

- Unutilized or underutilized employee knowledge and creativity

---

[7] www.asme.org/topics-resources/content/5-lean-principles-every-should-know.

Lean software development can be viewed as the application of the previous concepts and principles to the practice of developing software. Thus, the main aim and objective of Lean software development is to enhance customer value within budget and in the shortest possible delivery time.

■ **Pause and Consider**   Lean was created for manufacturing, but has evolved into a popular methodology for software development. How might it work in your organization? What do you see as the possible advantages? Do you think senior management might be impressed by some of the potential benefits?

Overlap with Agile Methodology

We would argue that Lean and Agile are two names for the same methodology. Looking closely at Agile software development practices, it is not too hard to notice close similarities with Lean concepts.

For instance:

- Code inspections and built-in quality in Agile support one of the cornerstones of what Lean methodology is founded on. The principle is not to push poor-quality to downstream processes; it means to find and fix defects early in the process flow and not to find them when the whole product is ready to be released.

- Scrum uses a Lean "pull" principle to balance a workload and to achieve a smooth flow through the system. The team pulls stories from the list in each Sprint (iteration) based on team capacity and capability.

- Agile (Scrum), like Lean, focuses on waste elimination by creating processes to identify and eliminate waste from the software development cycle.

- Agile uses Kanban techniques to help manage the development of the product and to increase communication between team and stakeholders.

- Agile rituals like morning standup, planning session, or retrospective session, all trace their origins to Lean.

We observed one team that faced heavy application-support tasks for multiple internal customers. The team used the Scrum framework for development of new features and solutions, but Kanban was used for tactical ad hoc-based support to fix breaks and address issues with their products. The team started with a simple pencil-and-paper board to track items on the fly and to track new, incoming requests. In this particular situation, there was a need to bring more visibility and workload management to the team. With time, the board evolved, bringing more complexity, with the ability to track the team's performance and make it easier to identify and quantify the waste, especially in the area of waiting for items to develop and in the area of rework. The board was moved from pencil-paper based to a digital format and then was split into two different functions, one for the development of new products and services and the other to support the existing released solutions. The board evolved together with the team's growing maturity in the Agile methodology.

There is no doubt that Agile methods and Lean software development go hand in hand in creating value for the customer. However, we believe that there are key differences between Lean and Agile.

Agile seems more tactical in nature where much focus is given to the technical practices, rituals, and techniques for effective and efficient software development.

In our opinion, little is said about what enterprise-wide operating framework and procedures must be in place to make Agile adoption successful. Therefore, to become an Agile enterprise organization, the major changes required must be initiated and driven from the top. H. Smits (2007) states that the "experience gathered during large scale implementation of Agile concepts in software development projects, teaches us that the currently popular Agile software development methods (like Scrum) do not scale to program, product and organization level without change. The fundamentals for changes to these methods are found in Lean principles...."[8] At this point, an organizational strategy becomes the main driving force within which Agile processes could operate effectively.

Instead, Lean is applied from an upper management perspective, with the objective of optimizing all aspects of their activities across the entire organization.

From this viewpoint, Lean could be seen as a top-down approach while Agile is bottom up.

Without the strategic piece, Agile adoption is depressed by the organizational forces that are resistant to change. By having Lean introduced from the top down, the tension between developers working in an Agile environment and their stakeholders due to conflict of practices and principles would ease.

---

[8] Smits, H., 2007. The Impact of Scaling on Planning Activities in an Agile Software Development Center, in: Proceedings of the 40th Hawaii International Conference on System Sciences (HICSS'07), Waikoloa.

We believe that the Agile framework does not generally concern itself with the surrounding business environment in which the software development is taking place. It primarily provides a set of practices designed for use by developers and can be viewed as supportive to Lean methodology. Lean principles, on the other hand, can be applied to any scope, from the specific practice of developing software to the entire organization in which software development is just one small part.[9]

Poppendieck and Poppendieck (2003, 2006) viewed Lean principles as core truth that do not change over time, while software development tools and practices are the application of those principles to a particular situation. Those tools and practices should be tailored accordingly, depending on the environment, and change as a situation evolves. Consequently, they suggest that Lean thinking should be viewed as guiding principles to develop and adapt Agile practices.[10]

Starting down the Lean/Agile road can be difficult. We believe it would be very useful to look at developing an adoption or scale-up roadmap for companies to trial Lean and Agile practices within their business environment while minimizing operational risk. Just keep in mind that by creating a Lean/Agile island in the organization would not bring a huge breakthrough. However, it would bring enough evidence to learn while scaling up across the organization. Remember, Toyota spent 30 years to refine their production system; it is a journey that would require persistence and lots of effort.

■ **Pause and Consider**   There is a lot to consider between Agile and Lean and a lot of crossover. How do you think either would fit in your organization? Who in your organization could you talk to about Agile and Lean principles, to get additional input?

DevOps

The term DevOps combines development and operations practices and tools into one framework, which is designed to increase an organization's ability to define, develop, and deploy IT solutions and services faster than traditional software development methodologies would do.

---

[9] For additional information on this topic, please see: https://citeseerx.ist.psu. edu/viewdoc/download?doi=10.1.1.905.3689&rep=rep1&type=pdf.
[10] Poppendieck M, Poppendieck T, 2003, *Lean Software Development: An Agile Toolkit for Software Development Managers.*

The core principle of DevOps lies in removing barriers between siloed development and operations teams, to bring their work into one framework to work together throughout the entire IT solution life cycle, from development and testing to deployment and operations. In other words, DevOps brings people, process, and technology together to constantly deliver value to the customer. In that framework, both teams communicate frequently by setting strong cultural ethics around information sharing through the set of chats, recurring design, problem-solving sessions, and huddles. This creates an environment in which communication flows rapidly across all engaged stakeholders such as developers, operations, and supporting teams.

■ **Tip** Silos sometimes occur because of individuals or managers "hoarding" information. Tread softly as you work to eliminate them.

To avoid the long solution deployment lead times, developers work on small increments that are released independently of each other. To achieve that, DevOps uses continuous integration (CI) and continuous delivery (CD) pipelines in combination with automation to move code from one step of development and deployment to another. For CI, the key goal is to find and fix bugs as soon as possible to improve quality and time of delivery. The CD practice ensures that workable software can be built in short periods of time and is ready to be reliably released at any time without manual intervention. To deliver software quickly, DevOps tries to rely on automation tools as much as possible. A great level of automation would ensure testing and preparation for release to production with higher speed and frequency. When CD is properly implemented, developers would always have available built, tested, and ready-to-production items.

Because of the continuous nature of DevOps processes, it is easier to visualize them as an infinity loop (Figure 9-1). The loop consists of the phases on the development and the operations sides and represents the relation to each other throughout the DevOps life cycle.

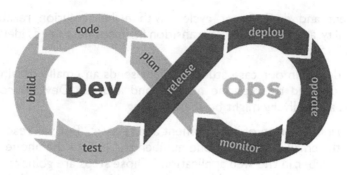

Figure 9-1. DevOps infinity loop[11]

DevOps can perfectly coexist with Agile software development and is a great framework for large IT-driven transformative projects to meet business needs. DevOps can address the challenges that limit business growth, including delays in the release of new software; sometimes the delay is so long that by the time it's released, it doesn't meet business needs. In today's fast-paced environment, where technology changes occur at a rapid pace, an organization whose revenue is heavily dependent on IT capabilities must be flexible and be able to adjust quickly. Time is critical and any delays may shift an organization's market position from leader to follower in a matter of a few months.

Therefore, the main benefits of DevOps is in

- — Higher speed of innovation.

- — Better adaptiveness to constantly changing market needs.

- — Rapid value delivery to build or maintain a competitive advantage.

- — Higher product/service reliability, much easier to ensure quality within smaller releases.

- — Better collaboration between technology and operations; development and operations work as one team by combining their workflows and by sharing accountabilities.

In addition, DevOps becomes popular among Lean and Agile practitioners and continues to evolve as artificial-intelligence-powered solutions come to aid in almost everything we do, from reviewing inquiries, searching for and analyzing data, performing digital tasks, and proposing decisions to incident management. It means smarter solutions would require even shorter

---

[11]www.pikpng.com/downpngs/imwwoJh_development-wants-rapid-change-operations-wants-stability-devops/.

development and deployment cycles, with higher solution reliability and built-in quality, and with seamless transition from business needs identification to viable IT solution.

Let's consider again our case study. A client sends an email notifying us that they have changed their phone number and address. DevOps continuous integration and delivery might be the following:

- First iteration of development and release: "Bot will read the email, read the phone number, and update the phone number in the web application." Those steps are going to be automated; the address changes on the web application and mainframe and phone changes on mainframe will still be processed manually. The business will be asked to approve and accept this first iteration.

- Second iteration: "Bot will read the email, read the address, and update the address in the web application." The mainframe phone and address change will still be updated manually. The business will be asked to approve and accept this second iteration.

- Third iteration: "Bot will read the email, read the phone, and update the phone number in the mainframe system the address change on mainframe will still be changed manually."

- Fourth iteration: "Bot will read the email, read the phone number and address, and update the phone number and address in the mainframe system."

With each new iteration, you are increasing the scope of automation. Without lengthy requirement gathering, the steps of approval, RPA development, and testing the solution move from conceptual solution to live solution faster. Those shorter cycles could prevent shifting requirements and solution scope creep, both of which are common when teams are working on complex and large solutions. This incremental and continuous delivery approach may be a new concept for the business, which has been accustomed to only approving the entire requirements document, testing, and accepting end-to-end automation at once.

■ **Pause and Consider** Think about hybrid versions. For example, RPA may do development, but some other team may gather requirements, do the deployment, etc. Would this be practical in your organization? If it is required, what steps will you take to help ensure success?

Regardless of what methodology you select, the main challenge for RPA is that, most of the time, the teams are new in the organization and usually follow an Agile or Lean methodology. Existing business partners and tech teams may still be in Waterfall mode, or in transformation, and Agile or Lean, with their own vocabulary, practices, rules, guidelines, etc., may be very foreign, and thus intimidating, to them. How do you build bridges between existing methodologies and the new methodologies that RPA will introduce?

We discussed four methodologies earlier in this chapter. Now we need to see how to incorporate the new RPA methodologies into whatever the existing corporate methodologies are.

First, let's look at some of the challenges:

- The RPA team will be discussing Sprints, backlog items, product increments, and other terms that the business may not be at all familiar with.

- The RPA team will present the business with a very limited requirements document, the MVP (minimum viable product), as discussed earlier. But the business may be expecting the complete requirements.

- The business may allocate a resource to review the requirements a month or two after the project starts, depending on how complex it is, with the belief that it will take that long to document the requirements. Yet that resource will be needed throughout most of the project life cycle.

Let's consider that RPA teams are more role based than team based. By that, we mean that while there is a core group of developers, a business analyst and a system analyst (other or different roles depending on the organization), there are other roles that will work on specific RPA projects, but not others. For example, in our case study, we would need someone from customer service to be part of the team from start to finish. That person (or persons) would help draft requirements in an interactive manner, and so would be needed throughout most of the project. For a different project, perhaps one that assures that taxes on merchandise adhere to government regulations, no one from customer service would be needed, but someone from the company's regulatory team would be.

So with this model, the permanent team members (e.g. developer, process analyst) remain on that team; other needed resources are shared as required. After the project completes, the shared resources are released back to their original team.

■ **Tip** Shared resources may not be required full time on the team. Determine the need on a project-by-project basis.

One of the potential downfalls is that, after the shared resources return full-time to their original teams, they won't be available to address issues that arise. You must plan for how you will handle issues that those resources were expert in.

Pause and Consider   What specific steps do you need to take in your organization to determine the best methodology as you introduce RPA? Who do you need to confer with? Who needs to buy in? How will you get that buy-in?

As indicated, there are a variety of development methodologies for you to choose from. But you needn't be overwhelmed; review what is current in your organization, determine which methodology you think would be most successful with your new RPA initiative, and work to implement it. But remember, what may appear to have great potential for your organization may wind up having to be changed. That is fine! Ideally, RPA is iterative; you learn as you go along. If something doesn't work, understand why, and either remove the roadblocks, or go in a different direction.

In the next chapter, we will discuss how RPA is constantly evolving and provide details on the most important trends for you to consider.

Planning for the Future

While RPA is relatively new, it is sweeping through technology companies, due to the need to automate repetitive, manual processes. But as technology is ever changing, RPA is evolving with it. It is important for you to know about coming trends, so you can determine if and when to introduce them into your organization. We will discuss the most significant "evolutions" to RPA here.

RPAAS, process mining, and intelligent processing are extensions of RPA.

RPAAS – RPA as a Service

Currently, most companies have native clients for automation tools. In the future, and the transition is already happening, all infrastructure of RPA is moving to the cloud. As stated in *Robocloud*, "Using RPA as a service means that your company does not have to purchase your own servers, licensing or professional services."[1]

---

[1] https://robocloud.co.uk/rpa-as-a-service/.

© Robert Fantina, Andriy Storozhuk, Kamal Goyal 2022
R. Fantina et al., *Introducing Robotic Process Automation to Your Organization*,
https://doi.org/10.1007/978-1-4842-7416-3_10

The basic advantage is that all the benefits of cloud can be leveraged. These include

- Increased capacity: For example, perhaps you have ten Bots running on one machine. If you need to add more, you will need to increase the memory, the processing speed, etc. With RPAAS, you have sufficient flexibility to increase the capacity and replicate servers much more quickly. You can replicate the machines/servers on which the Bots are running, enabling your working force to be replicated faster.

- Increased flexibility: For example, we know that Bot licenses can be expensive. Currently, if you need 2 Bots immediately, and you estimate that within the next 6–12 months, you may need 2 more, you need to buy 12 licenses and create the infrastructure for them. With RPAAS, you take advantage of the cloud and you can replicate machines in minutes, so you can increase or decrease the number of Bots at any time. By working with the business to develop a sound business and technology strategy, you have the potential to realize significant savings.

- Sometimes new functionality is needed very quickly. Using RPAAS, you can adapt new releases and features quickly, because you can scale up the infrastructure on demand.

Pause and Consider   Based on what you now know about RPA, how do you think RPAAS could benefit your organization? What steps would you need to take (e.g., a clear presentation of potential cost savings, based on historical information) to convince management to take this next step?

Process Mining

We have emphasized throughout this book that RPA is for repeated processes which have simple logic and don't need much intelligence or human intervention. This is because with most processes, there are security concerns and concerns about accuracy. The process has to be simple enough that security and accuracy standards are not compromised. In our experience, processes are introduced to us by business leaders who want to streamline their operations and improve efficiency. So they may identify one or two processes that they know are extremely repetitive and would appear to be good candidates for automation. But often these are processes which for some reason are getting some attention; perhaps the people performing the process are voicing discontent because, we'll just say it, doing this kind of repetitive work can be boring!

There may be several other processes within the business unit that are just as repetitive, but perhaps they aren't coming to the notice of management at the moment. Perhaps they are month-end or year-end, rather than daily, processes. Perhaps they are only needed occasionally to clean up some kind of backlog of requests. Regardless of the reason, they may be excellent candidates for automation, but are not currently identified as such by management.

Process mining tools have been designed to analyze processes, and using built-in logic, identify and help determine which are the best candidates for automation. The business provides as many processes as possible, not just those that are the "squeaky wheel" processes, and enters them into the process mining tool, and the tool will identify the best automation candidates.

At a simpler level, with smaller organizations or simpler processes, process mining tools can also assist as a starting point for identifying automation candidates. The tool can highlight significant business functions and their accompanying processes.

Another area where process mining tools can be highly beneficial is in the meeting points between processes. Process mining tools can assist in determining where those intersection points can be automated.

Thomas H. Davenport and Andrew Spanyi, writing in the *Harvard Business Review*, sum up the advantages of process mining succinctly:

> *Process mining software can help organizations easily capture information from enterprise transaction systems and provides detailed — and data-driven — information about how key processes are performing. It creates event logs as work is done: an order is received, a product is delivered, a payment is made. The logs make visible how computer-mediated work is really happening, including who did it, how long it takes, and how it departs from the average. Process analytics create key performance indicators for the process, which enables a company to focus on the priority steps to improve.[2]*

Remember, due diligence is still required; the tool is only as good as its input, and the automation decision must still be made by a human resource.

Once processes are selected by the process mining tool, all the phases from "opportunity assessment" shown in this book would be followed, including periodic meetings with the Governance Committee to approve or disapprove moving to the next step.

---

[2] https://hbr.org/2019/04/what-process-mining-is-and-why-companies-should-do-it.

■ **Pause and Consider**   For larger organizations with highly complex processes, process mining can greatly expedite the identification of potential automation candidates. Would this be beneficial in your organization? Who would you need to speak with to better understand the potential benefits of process mining to your organization?

Intelligent Processing

Currently in RPA, we build the logic and simple rules for converting the manual steps into automation. Intelligent processing can be viewed as "a way to process unstructured textual data. Document cognition leverages artificial intelligence, machine learning, cognitive science, and natural language processing to index structured, semi-structured, and unstructured data."[3]

With intelligent processing, we are building IQ Bots. Using artificial intelligence and machine learning, along with RPA techniques, we are adding intelligence to the rules when building Bots. One example is identifying handwritten text in documents; currently, information input to a Bot with handwritten information will be diverted for manual handling; the Bot cannot read handwriting. AI techniques can assist in reading this kind of text and have a percent of accuracy that can be calibrated so the performance of the Bot can be increased. As with any process that is partially automated, anything the Bot cannot read and/or understand can be diverted to manual processing.

Writing in *Capacity*, Joe Sullivan listed seven benefits of intelligent document processing.[4] He said the IDP

1) Promotes automation because it facilitates process improvement

2) Simplifies compliance by sorting documents, data entry, and information validation

3) Is effective, because it reduces manual intervention in document-centric workflows

4) Supports scalability, because it can be applied to various applications in different areas in your organization

5) Attracts savings by reducing the duration it takes to process documents and reduce the cost of labor

---

[3] https://thechatbot.net/intelligent-content-processing-for-chatbots/.
[4] https://capacity.com/seven-benefits-of-intelligent-document-processing-idp/.

6) Increases speed by processing large volumes of data in a very short amount of time

7) Enhances customer experience and satisfaction by greatly reducing the time it takes to process paperwork

---

■ **Tip** As Sullivan mentions in his article,[5] some management teams tend to overlook the soft costs related to customer experience. Don't make that mistake! Just because you may not be able to quantify customer experience in terms of dollars and cents, any organization must remain laser focused on customer experience if it wants to succeed in today's market.

Hyperautomation

One of the upcoming technological trends which is heavily discussed in the industry today is hyperautomation. According to Gartner,[6] hyperautomation in the near future, within the next few decades, is going to have the greatest impact on expanding automation of business processes.

As we have already mentioned earlier, the best use of RPA is for highly repeatable, high-volume, manual tasks carried out by employees. Unfortunately, traditional RPA software has limitations. It doesn't understand inputs in the form of unstructured data, and it doesn't learn by itself. Due to those limitations, any RPA solution is only suitable for rule-based tasks which require human intervention to translate them into the Bot's language, in the form of code. It leads to a few disadvantages. First, any changes to the process would require a skilled developer to make changes to the code and then test and deploy the changes to production. It slows business agility and increases operational costs. Second, depending on the industry, the rule-based processes may cover a small fraction of overall business operations, limiting the expansion of digital processing and not achieving operational cost reduction targets.

So, how about tasks that require more critical thinking? What if most of the processes in your organization require analysis of different data to produce an output? Can they be automated?

The concept of hyperautomation goes beyond the straightforward, rule-driven, and repeatable processing; it is not a single tool but a combination of, in addition to RPA, multiple, more advanced processing technologies, such as intelligent document processing (IDP), optical character recognition (OCR), and intelligent character recognition (ICR). It combines AI-powered solutions

---

[5] Ibid.
[6] `www.gartner.com/doc/reprints?id=1-253WYNPK&ct=210129&st=sb`.

like predictive modeling, machine learning, deep learning, natural language processing, and other intelligent processing technologies into one system. Compared to conventional RPA automation, hyperautomation allows expanding automation to complex and often heavily cognitive processes that require human intervention to make a decision, enabling end-to-end process automation of the business operations. Its primary mission is to continuously increase the scale of automated end-to-end operational decision-making processes by eliminating human operators from low-value operations. Due to its nature, hyperautomation creates a separate subset of digitally run operations within an organization that replace conventional operations that rely heavily on manual processing, often requiring the physical presence of the workforce.

Consequently, in the near future, operational leaders will need to be ready to redefine their employees' responsibilities, their required skill set, and the way they are seeing their work. It might seem that the expansion of hyperautomation across industries would make human employees obsolete. We would argue that there is no reason to be so pessimistic, at least not for a very long time. We believe that the artificial intelligence solutions as part of hyperautomation will redefine and reshape the workforce, and, yes, some jobs are going to be fully automated, but some new ones will be created and others will need to be readjusted. Looking at the nature and limitations of AI-powered solutions, the jobs that will likely disappear in the future are customer service related such as receptionist, travel agent, cashier, bank teller, etc. We would predict that manual processing would still exist in the future for very complex decision-making operations. However, we envision that AI will function more as an assistant to human decision-making and not its main driver. In those human-machine models, AI would process input data and make recommendations, or propose a decision together with the rationale behind it, but a human is going to be the one who makes the final call.

An extension to this AI-human partnership would be an incorporation of a learning loop back to AI when a human makes a decision to create a continuous learning cycle. The same way as we humans learn by gaining experience with time, machines will learn from humans. The proposed approach, in comparison with traditional supervised algorithm-based training, would accelerate the AI learning process, but most importantly, it increases the adaptability and speed with which any change in the decision-making model can be incorporated. As business conditions are rapidly changing around us, AI will need to be kept updated with current criteria and business rules in order to avoid a model to drift and potentially result in inadequate performance and poor decision outcomes. Thus, establishing continuous learning cycles where humans continuously curate the decision criteria back to the AI model would ensure continuity of business operations by avoiding taking AI out of operations for lengthy overhauls and adjustments.

Further, depending on risk and poor decision impact level, the human-AI model can be adjusted so that high-confidence decisions could be delegated to machines, allowing humans to focus on low-confidence decisions. Of course, the question is what parameters to use and how to define the confidence level based on which criteria.

Hyperautomation could bring numerous advantages that could boost an organization's performance, increase its competitiveness, accelerate innovation, as well as improve the well-being of its workforce.

These could include

- *Reduction in operational cost* by redesigning and replacing manual, low-complex processes with automation.

- *Acceleration of digital transformation* when the organization could align its business operational goals with investment in technology.

- *Improved competitive advantage* by introducing new technologies into day-to-day operations, since outputs could be produced faster, with higher quality and lower cost, resulting in improved customer satisfaction.

- *Increased quality of business decisions*: AI technology would require access to data; this means that business- related information can be easily extracted and used, which would create an effective evidence-based, decision-making environment.

- *AI-powered technologies would accelerate innovation* across the organization and increase business agility.

- *Improved collaboration among technology and business operation teams*. Hyperautomation requires integration of multiple technologies and business operations, which are traditionally siloed, into one ecosystem.

- *Increased employee satisfaction* since it would enable employees to focus more on value-added tasks which are going to be more interesting, challenging, and diverse. Creating a smart working environment where employees will be challenged and do not waste their time on boring, repeatable, low cognitive, and low value tasks would result in an increase of their productivity

Hyperautomation can be largely advantageous for large and hierarchical organizations with many legacy systems and heavily manual operations that have considerably low automated processing levels. RPA Bots can integrate legacy systems with other technology solutions used in hyperautomation.

By achieving that, those organizations would see real results in improving business operations' agility, effectiveness, and efficiency very quickly.

It is important to remember that different technologies used as part of hyperautomation might experience challenges during their implementation due to organizational and technological limitations. The most common challenges include

- Using AI-powered cloud-based solutions may require the use of client personal data which could increase risk of a privacy breach.

- AI-powered solutions may inherit biases that are either coming from the data used for training or embedded into an algorithm because of using assumptions.

- Applying hyperautomation to poorly defined and understood processes. It is difficult and pointless to automate poorly documented and unclear processes. Process mining tools earlier mentioned in this chapter can help to bring visibility and better understanding and assessment of the processes.

- Applying hyperautomation to serve customers with extremely complex and unique requirements would require a craftsmanship solution. If process steps and outcomes are customized for specific customer needs to increase customer satisfaction, it will introduce complexity and dependency on humans for manual handling. It is very important for any organization to define the right balance between customer satisfaction and process simplification through its standardization.

- A risk-averse organizational culture, which promotes avoidance of potential errors creating conservative inertia, would slow down innovation and adoption of automation. Empowered employees should be able to explore, experiment with new upcoming automation technologies, and foresee business benefits quickly. It is critical to ensure end-to-end, technology, and business operations leadership support, gain their trust, and confer with them to try new and unfamiliar approaches.

- Selecting the right stack of automation solutions from constantly changing and evolving products available on the market. We believe that an organization should embrace automation as an operational principle rather than simply seeing it as added functionality or tool.

We see that in the near future, RPA capability will be a part of the automation stack of products that are available in out-of-the-box functionality of workflow software. There is going to be no need to purchase it separately and integrate it to the tech architecture.

Now that you have all that you need to embark on your RPA journey, you need to be aware of some of the challenges you may encounter, beyond those already mentioned. Chapter 11 will help you avoid some of the issues we have encountered in our experience.

Challenges and Pitfalls

Once you have created the foundation for your RPA initiative and have identified a few "early adopters" who are willing and anxious to have some of their processes automated, and have established your governance policy and identified board members, it's time to look ahead and see what challenges you may encounter. You can minimize some common mistakes by understanding what we did initially that wasn't successful.

Throughout this book, we have mentioned many of the possible challenges you may encounter when starting your RPA journey. Some of these we were able to avoid, seeing the speeding bullet and figuring out how to dodge it. Some we crashed into head-on and then resolved the issue after the fact.

Studying these will assist you in being able to proceed smoothly as you work to automate manual processes in your organization:

- Expecting easy management acceptance. When faced with an opportunity to cut costs and improve efficiency, we thought management at least would welcome the new RPA program. We were mistaken. While line workers were sometimes reluctant, due to a fear of job loss, we were surprised by more than a little management resistance. Some managers seemed to think that RPA

© Robert Fantina, Andriy Storozhuk, Kamal Goyal 2022
R. Fantina et al., *Introducing Robotic Process Automation to Your Organization*,
https://doi.org/10.1007/978-1-4842-7416-3_11

was the new buzz phrase, the solution of the moment, which would generate much excitement and quickly and quietly fade away. They had seen too many other "game-changing" solutions that had done just that. Others seemed concerned not for their own jobs, but for threats to their "empire"; they had X number of employees and didn't want that number reduced. If a Bot was going to do some of the work their employees were now doing, they would possibly lose those employees, since they could be repurposed elsewhere in the organization or let go.

Tip If you are careful to follow the information in Chapter 2, you will be able to avoid this challenge.

- Identifying the wrong processes to automate. It may seem simple to identify the right process; process XYZ is repetitive, so let's automate it. But on closer scrutiny, it wasn't as simple as we thought. As mentioned earlier, when you first identify a process for automation, you don't know too much about it. But there are some "red flags" that we missed in some of our first attempts. For example, is a middle-management employee deeply invested in that process as it is currently performed? Would automating a process possibly step on the toes of a manager who does not want to recognize that their area could be improved? Or is it something as simple as an extremely complex process that could have been determined in a preliminary conversation? We learned in time to better discern these issues.

- Attempting to automate an entire process when only part of it is a good candidate for automation. This is related to the item just mentioned. A process with multiple steps may be very manual and very repetitive, but that doesn't mean that all of the steps can be automated at this time. The business owner may envision having the entire process automated, but that isn't always reasonable. Remember, you may be able to automate a part of a process and still achieve significant savings in time or cost or provide a greatly enhanced customer experience. Look at the end-to-end process as you have mapped it, consider the time savings, and if only some of it can be automated, and that will achieve good results, then automate just that part of it. Work with the business owner to determine what may need to be done to make additional steps in the process better candidates for automation in the future.

- Attempting to build the Bot without the required expertise. The developers who will do the actual work of building the Bot must have the knowledge to do so. Enthusiasm is great, but it isn't a substitute for experience. Be sure to have at least one developer on the team who is highly qualified in RPA. They can mentor and train less-experienced developers.

Pause and Consider   What potential "land mines" can you foresee in your organization? Who are the people who are particularly protective of their areas? How would you approach them if you are aware of process opportunities in their areas?

- Governance board meetings. Initially, we submitted the document package with the intake form, business case, risk assessment, and other documents several days prior to the meeting to the governing board members. They and the requestor(s) were invited to the weekly meeting to review the request.

- We found that, generally, no one read any of the documentation prior to the meeting. Therefore, instead of sending out detailed information in advance, we summarized the information into a page or two which we presented at the meeting. We stopped sending the detailed documentation, but maintained it for our records and just provided a high-level overview of the request at the start of the meeting. This overview consisted of the following:

    - A brief description of the request (e.g., automate process XYZ which currently takes three people two days every week to accomplish).

    - A review of the savings (e.g., this will free up 2,250 hours per year [3 people times 15 hours per week times 50 weeks per year]). Multiply this by your company's hourly rate.

    - An evaluation of the complexity of automating the process (e.g., high, medium, low).

    - An estimate of how long development will take (e.g., three weeks).

The architect who worked on the initial estimates should be present to answer questions.

- Inviting the requestor to the governance board meetings. We initially thought there was an advantage to this, in that the requestor could answer some questions about the request that the governing board members might have. But those questions were able to be answered by the business process analyst or the architect, and if the request was not approved to move forward, it didn't provide anyone with the opportunity for tactfully explaining this to the requestor. Now if a request is turned down by the governing board, the requestor will receive an email similar to the following:

  - "Hello (name),

    Thank you for your recent request to have process ABC automated. We have reviewed your request and evaluated the costs of automation, savings to be accrued, urgency of need, and other factors. At this time, there are higher-priority requests that we must focus on. Therefore, your request will be placed on hold to be periodically re-evaluated.

    If you are aware of other processes that could possibly be automated with this one, please advise us.

    If you have any questions about this, please contact me.

    Sincerely,

    (Name)

    (Title)"

In the governing board meeting, the discussion is generally more direct, with board members possibly saying the process isn't worth automating. This discourages the requestor from submitting other requests and casts a negative shadow on RPA.

Sometimes a process isn't right for automation, but there may be other areas of the company that could provide some assistance. In those cases, that information should be included in the email to the requestor, and the RPA team should engage that other area and refer the request to it. No further work, except for the periodic re-evaluation of all projects on hold, is required for this particular request.

■ **Pause and Consider**   How would you best advise a business process owner who requested automation of a process that that request was denied? Remember, you don't want to leave them with the impression that their request itself was not valid, only that the numbers didn't work well enough to proceed.

- Not obtaining sufficient information early on. Sometimes, even when we worked with the requestors in completing the request form, there was information missing. We adapted the form going forward and eliminated that issue, but be aware that the requestor may not have all the necessary information. That is why it is vital to meet with subject matter experts, the people who know how the process actually works, not how the documentation about it says it should work. There is often a significant disconnect between the two.

■ **Tip**   Remember, the templates are made to be tailored to your specific needs. You may want to use them "as is" initially and see what works and what doesn't work for your organization.

- Not adequately addressing concerns about job loss due to automation. Sometimes this could be a real fear; management may see headcount savings as an advantage to automation. No one wants to assist someone in causing them to lose their own job.

■ **Pause and Consider**   It's possible that the goal of any particular automation may be to reduce headcount. How will you deal with this? How will you get assistance and information about a process from people who may be fully aware that by automating the work they do, they will be out of a job? What assistance and support can you get from your HR organization or senior managers?

- Trusting existing (old) documentation. When a new process is introduced, there is often accompanying documentation. This documentation explains the purpose of the process and, more importantly, how the process is supposed to work. However, there is often a major disconnect between how the process is *supposed* to work and how it *actually* works in day-to-day practice. Sometimes when a request for automation is received, that documentation will be provided. But it is vitally important to map the process with the SMEs, the people who are actually doing it. A process may work very differently from the way it is supposed to work, and the longer the process exists, the farther it may be from what was initially documented. Be sure to map the process *as it actually is performed today*.

■ **Tip** Process mapping with the SMEs will assure that you understand the way the process actually works. Do not rely on any existing documentation.

- Selecting the wrong tool. There are many things to consider when selecting your RPA tool, including how much programming it requires to meet your needs, how scalable it is, what is its record of changes and promptness in providing updated features, and how much maintenance does it require? And a big one for management, what is the cost? Just because a tool is expensive doesn't mean it's the best. Do your research: look online, talk to industry peers, and review journals to find out what RPA tools would work best for you. Remember, a tool that is ideal for one organization isn't necessarily the best one for every organization. Do your homework and avoid this pitfall.

- Overconfidence of the business. Once the Bot is up and running, there is always the risk that it will fail at some point. In that event, there must be personnel ready and sufficiently knowledgeable to manually perform the process until the Bot is once again functioning. Generally, there are one or two people assigned to handle the exceptions that the Bot can't handle (such as invalid client IDs as indicated in our case study), and they are usually the people who would take over in the event that the Bot stops functioning. But we found that this sometimes needs to be reiterated to our business partners.

This is not a comprehensive list of everything that could possibly go wrong. You will probably add to it with your own mistakes! But these are some key challenges that we either encountered or anticipated, and it's important that you be aware of them.

Summary

Congratulations! You now have not only the basic understanding of RPA and what it might mean to your organization, but you also have the details you need to get started and see the initiative succeed.

To summarize, first, what is Robotic Process Automation (RPA)? It is simply automating repetitive manual processes. Every organization has them. Perhaps it's a manufacturing company that bills for its products. The salesperson sells the item and enters the order into the system. A clerk finds the name and price of the item based on some serial or other identification number, looks up the name and address of the customer, types the invoice, places it in an envelope, and puts it in their outbox. They then get the next order and follow the same process, day in and day out. A "Bot" (robot) could do this: read the order, look up the customer's name and address, populate the fields on a blank invoice, and send it to the printer which may be equipped to fold invoices and insert them into envelopes. Then, once or twice a day, someone from the mailroom can pick up the invoices and send them. The clerk or clerks who were responsible for these tedious tasks are then freed to work on other tasks that actually take thought. Chapter 1 provided you with an excellent understanding of RPA.

If you have identified the need to automate processes within your organization, how will you get buy-in from senior management and line workers? If things are progressing well, senior management may not want to change. And line workers may fear for the safety of their jobs (this could be a real concern if one of the goals of automation is headcount reduction). By reading Chapter 2, you now know how to address these issues.

© Robert Fantina, Andriy Storozhuk, Kamal Goyal 2022
R. Fantina et al., *Introducing Robotic Process Automation to Your Organization*,
https://doi.org/10.1007/978-1-4842-7416-3_12

What should governance for an RPA program look like? How should it be structured? What roles should be involved? Who should sponsor the program? What will the sponsor's responsibilities be? And what overall framework will the program operate in? It must be structured, but be sufficiently flexible to adapt to changing needs. How is that done? As you read Chapter 3, you learned the answers to these questions.

Once you have senior- and middle-management approval and have identified some early adopters, how do you find processes for automation? What are the key components to look for that might make a manual process a good candidate for automation? As you learned in Chapter 4, a manual process that is basically done the same way repeatedly is one that should be evaluated for automation.

How do you solicit processes for possible automation? Once word of the RPA program becomes public in your organization, there may be many people who will be interested. That is excellent, but there must be some orderly manner for them to submit their requests. Each request must contain some basic information so you can make your initial evaluation. Chapter 4 discusses the request form, and there is a template for this in the Appendix (Appendix 1). Feel free to adapt it to the needs of your company, but try to keep it simple. You don't want people to feel intimidated by it.

When a manual process has been identified as possibly a good candidate for automation, you need to determine if, in fact, it is. You need to determine how much variation there is in the process, so you can know if it can be handled by a "Bot." How is this done? You also need to estimate what the savings will be in time (and time equals money, as we all know). After you determine that, you need to estimate what the cost of automation will be, so you can determine return on investment (ROI). And are there other considerations besides ROI? What if your major competitor is getting products to market faster than your company, and management sees some of your customers moving to the competition? How do you evaluate if that is worth the cost of automating some processes to be more competitive, even if the money saved by automation will be minimal? Chapter 5 provides you with the tools you need to make these decisions. It also helps to understand that while the initial assessment may indicate that the process is definitely worth automating (cost of automation will save money: good ROI, etc.), as you proceed through the life cycle of the individual automation project, additional information will be discovered. Some of that information may indicate that the cost of automation will be significantly higher than initially estimated, or the cost savings will be lower. At that point, a decision must be made (and here is one area where the sponsor plays a role) as to whether or not to proceed with the project or cut your losses and cancel it. Doing so is NOT a failure; our RPA program is designed for incremental steps and corresponding milestones to evaluate the information and work completed in each phase. Excellent work may have been done during *opportunity assessment*, but as developers and others work through *solution design* (Chapter 6), they will uncover more information, and some of it may indicate unanticipated expenses in development or less savings than initially estimated. It is better to find this information now, than to proceed through the expense of the entire project, go over budget, and find that the solution does not deliver as anticipated.

What happens if, after obtaining steering team approval to proceed to the next step, you learn, in that new step, that complexity is greater than initially thought, or the savings will be far less than estimated? You know from Chapter 5 that this simply indicates that the RPA program is working as it should. It is incremental: you gain some initial information and make a decision about moving to the next step. If that decision is made, you gain additional information and then make another decision about advancing. At any point, you may learn something that reduces the feasibility of completing the project. That is fine; this happens and is not a failure. It is far better to proceed a short distance into a project and then learn that some initial assumptions were inaccurate, than to spend the time and effort to complete the project and only learn at the conclusion that the expected benefit will not materialize or the cost of creating the solution far exceeded the amount the solution will save.

Who (what roles) needs to be involved in solution design? What specific responsibilities will they have? Chapter 6 discusses the following roles and their corresponding responsibilities:

- RPA technical delivery lead

- Architect

- Developers

As discussed in that chapter, there may be other roles that are required, depending on various circumstances in your organization. Tips on how to determine if there are, and what their responsibilities would be, are detailed.

Once you have your team established, and have identified some good candidates for automation, and designed the new Bot, you need to deploy it. There are many components to Bot deployment that may differ from the way you generally deploy new software. Chapter 7 details this information for you.

In Chapter 3, we introduced the centralized, decentralized, and hub-and-spoke models. In Chapter 8, we detail these models, with key information on their advantages and disadvantages to assist you in selecting the one most appropriate for your organization.

Today, there are many development methodologies: Waterfall, Agile, hybrids of those, Lean, DevOps, and others. All of them can be used effectively with an RPA program. There are some commonalities and some differences depending on methodology, and Chapter 9 provides the detail required to assist you in deciding which methodology is best or, if one is mandated in your organization, how best to incorporate RPA within it. This chapter also helps you understand some of the roadblocks you will encounter, in situations where you may need to work to successfully interact with a business unit that is using and familiar with a different methodology than you have chosen for RPA.

One of the advantages of RPA is its scalability; as things change within your organization (e.g., new applications introduced; company grows), RPA can change right along with it. David Chappell, in his article, "Understanding RPA Scalability: The Blue Prism Example," describes RPA scalability this way:

"What is RPA scalability? One way to think about it is to focus on three aspects:

> *Handling increased load. This includes support for large numbers of RPA robots working together to carry out many instances of a business process. It also includes a way to easily change which business process each of your robots is executing, letting them work on different processes at different times.*

> *Expanding the scope of usage. This aspect of scalability means support for broadening how and where RPA is used in an organization. You might start by automating a process in one part of your business, for example, then expand by creating RPA solutions for processes in other business units.*

> *Increasing the breadth of access. Automated business processes often need to access new technologies, such as new applications or integration technologies. Your processes might also themselves need to be accessed by other software. These kinds of expanded access can be viewed through the lens of scalability, because both let your RPA solutions be used more broadly."* [1]

No methodology or technology remains the same over time; it either evolves or disappears from the scene, replaced by something better. RPA is no exception. In Chapter 10, we discuss the latest trends you need to be aware of as you introduce RPA to your program. You will probably start with the basics of RPA, but Chapter 10 presents you with a clear idea of where you can take the program in the future and how it may benefit your specific organization.

---

[1] www.blueprism.com/uploads/resources/white-papers/Understanding-RPA-Scalability-The-Blue-Prism-Example-1.0.pdf.

The introduction of any new methodology always has risks. In Chapter 11, we explain areas that we encountered that caused us to pivot from the road we thought we needed to take to one that was smoother and more effective. Things we thought were quite straightforward still caused some people to question the validity of RPA. This chapter, of course, doesn't list every challenge you may encounter, but by following its guidance, you will avoid some of the headaches we suffered through. Following the information in the entire book should prevent a few of your own headaches!

Finally, the book concludes with a detailed Appendix. Here you can find templates for all the forms described in earlier chapters, with complete instructions on how and when to use them, and how to complete them. You learned, of course, that these are excellent starting points for any new RPA initiative, but time and experience will help you to know if and how they should be tailored to suit the needs of your particular organization. We recommend, however, that you initially use them as presented herein.

To conclude, you now have at your fingertips everything you need to succeed with the introduction of a Robotic Process Automation initiative in your organization. We wish you every success as you embark on your RPA journey!

A

Appendix

Appendix 1: RPA Opportunity – Request Form

Appendix 2: Process Map Template

Appendix 3: Opportunity Assessment – End-to-End Process Steps

Appendix 4: Feasibility Assessment

Appendix 5: Risk Assessment Checklist

Appendix 6: Opportunity Brief

Appendix 7: Governance Committee Decision Form

Appendix 8: Solution Design Document

Appendix 9: Functional Requirements

Appendix 10: Nonfunctional Requirements

Appendix 11: Deployment and Release Setup Checklist

Appendix 12: Process Design Document

© Robert Fantina, Andriy Storozhuk, Kamal Goyal 2022
R. Fantina et al., *Introducing Robotic Process Automation to Your Organization*,
https://doi.org/10.1007/978-1-4842-7416-3

Appendix 1: RPA Opportunity – Request Form

Initiative Name	
Sponsor	
Champion	
How Does the Process Currently Work?	
What Is the Goal?	
Common Business Drivers/Business Criticality	

Impacted Business Areas		**Impacted Systems/Tools**

Opportunity Type	**Duration of the Solution?**	**Estimated Operational Benefits (FTE hours)**

Priority	**Desired Delivery Date**	

Form Completed Date:		**Prepared by:**	

The following information describes each field on the form:

- **Initiative Name:** This is generally the name of the process.

- **Sponsor:** This is the person, usually an executive leader, within whose budget the organization requesting the automation resides.

- **Champion:** This is the leader directly responsible for the process that is proposed to be automated. They will also be responsible for removing any roadblocks that may occur and allocating required resources that would be needed during the automation development, testing, and deployment phases.

- **How does the process currently work:** The current-state process is summarized in sufficient detail that anyone reading it has a comprehensive idea of the purpose, process flow, repetitive nature of the process, and the reason it is a candidate for automation.

- **What is the goal?:** This indicates what portion of the process, or the entire process, is proposed for automation. It briefly lists the benefits of automating the process.

- **Business criticality:** This section describes why this process must or should be automated at this time and how it is aligned to strategic business goals. Is there a competitive need? Is there exposure to high risks? Are clients complaining about something that would be resolved with this automation? And so on. Please be aware that some requestors will consider all their processes as "business critical," but a little discussion with them may indicate that that is not the case.

- **Common business drivers:** Some possible examples include, but are not limited to, the following:
  - Reduced cycle time
  - Improved consistency and quality of work
  - Allow staff to work on higher-value activities
  - Expedite satisfying governmental regulations
  - Maintain competitive advantage
  - Etc.

- **Impacted business areas:** This lists the business units that would be impacted by this automation.

- **Impacted systems/tools:** There may be specific systems that have been purchased or developed that will be impacted; these should be listed here. Also, any tools, including Excel, Outlook, Word, etc., that are impacted are listed here.

- **Opportunity type:** Note here if this is a one-time automation, possibly to clean up months or years of backlog, or if it will be ongoing.

- **Duration of the solution:** If the process is expected to be phased out within a certain time frame, or if the anticipated purchase of a new tool in, for example, 3Q2X, will resolve the issues to be addressed by automation, note it here.

- **Desired delivery date:** Indicate when the automation is optimally desired.

- **Estimated operational benefits:** This is generally the savings in time and cost, but not every requested automation has a significant time and cost savings. Some are required to meet regulatory or competitive requirements. Some are needed to reduce the risk of human error. If the automation will result in some time and cost savings or revenue gains, this should be explained in detail here.

For example:

- 5 staff members spend 3 hours a week on this process

- $5 \times 3 = 15$ hours per week

- 15 hours per week $\times$ 52 weeks = 780 hours per year

If a process is to be automated to save time and money, an annual savings of about 1,500 labor hours, or one full-time position, may be required to justify the solution development and maintenance cost. However, this will be different in different organizations, so it is important to establish the right criteria for yours.

- **Form completed date:** Self-explanatory.

- **Prepared by:** Include the names of all the people, and their roles, who had input into the request form.

Appendix 2: Process Map Template

"Process Name"

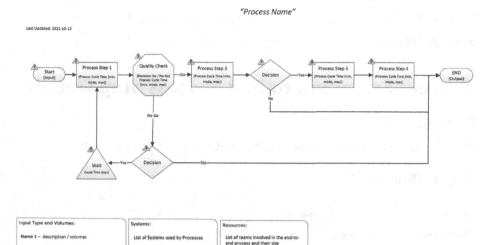

Last Updated: 2021-10-13

Appendix 3: Opportunity Assessment – End-to-End Process Steps

<Process Name>

Current State End-to-End Process Steps

Revision History

Date	Comment	Responsible Party
	<Brief description of update>	

Version: 00

21 May 2021

Contents

Step 1. [Name] _____ I

Step 2. [Name] _____ I

Step 3. [Name] _____ I

Current State End-to-End Process Step Details

Purpose:

<Briefly summarize the purpose of the process; what does it do?>

Step 1. (Name)

Input	Step Input Type(s)
Activities	• Sub-step 1 • Sub-step 2 • Sub-step 3
Output	• Output Type 1 • Output Type 2 • Output Type 3 • Output Type 4
Roles	
Metric	
Notes	

Step 2. (Name)

Input	Step Input Type(s)
Activities	• Sub-step 1 • Sub-step 2 • Sub-step 3
Output	• Output Type 1 • Output Type 2 • Output Type 3 • Output Type 4
Roles	Resource or Team Name
Metric	• Metric 1
Notes	Purpose of the step is to …

Step 3.

Input	Step Input Type(s)
Activities	• Sub-step 1 • Sub-step 2 • Sub-step 3
Output	• Output Type 1 • Output Type 2 • Output Type 3 • Output Type 4
Roles	Resource or Team Name
Metric	• Metric 1
Notes	Purpose of the step is to ...

END

****

Explanation of detail:

- **Step name:** This coincides with the information on the process map. If the first step is "Receive Customer Request," that will be the name you use on this template.

- **Input:** For the first step, this will describe what triggers the process. For example, to use our case study, the first input might be "Customer notifies company of changed address or phone number." After that, the "input" to a step is generally the "output" from the previous step.

- **Activities:** This is a description of what is actually performed at this point in the process. For the first step, to use our case study, it might be this: "The request is received by customer service. The representative reviews the request and forwards it to the appropriate department for handling."

- **Output:** This indicates what happens when the work in this particular step has been completed. In our example, the output might be this: "Request forwarded to Department ABC for processing." This will generally also be the input for the next step.

- **Roles:** Here, you put the name of the role, not the person, who handles this particular step.

- **Metric:** This is optional, but over time, you may find it quite useful. In our case study example, this might include "time it takes to determine the correct department to send the request." In that case, you might put "30 seconds" or a range, such as "30 to 60 seconds."

- **Notes:** This field is also optional, but can be used for any additional information you feel is required for the step to be understood.

Appendix 4: Feasibility Assessment

<Name of project>

	Questions	Answers
1	Is the process well defined?	
2	Is the process stable (very few "exceptions")?	
3	Can exceptions be handled manually?	
4	Are inputs in digital format?	
5	Can required data be input without human intervention?	
6	Are potential changes to roles and processes acceptable to management?	

Approvals:

<Name> <Role> _____ Date: _____

<Name> <Role> _____ Date: _____

<Name> <Role> _____ Date: _____

<Name> <Role> _____ Date: _____

Details of the form:

1. **Is the process well defined?** Here, you need to determine how repetitive the process is. Basically, are the exact same steps performed each time the process is run? Are the steps well-known? Are there a limited number of exceptions? This is one characteristic that makes a process a good candidate for automation.

2. **Is the process stable (very few "exceptions")?** When the process is invoked, are the steps that are taken based on clearly established rules? Are the steps the same within the process?

3. **Can exceptions be handled manually?** For any process, a certain, limited number of transactions might need some additional steps. Perhaps an account number was entered incorrectly, and the account must be looked up by last name. A customer might make a special request within a more common request. Can these transactions, if within an automated process, be sent to a mailbox for manual handling? Is there a process for handling these exceptions now?

4. **Are inputs in digital format?** Do the inputs currently come into the process via an electronic mailbox? Or are they calls from clients or others that must be input by the call receiver? If so, can the input created by the call receiver be input into a system that the Bot can access?

    Also, are all inputs Excel files, Word documents, PDFs, etc.? How easily might they be read by a Bot?

5. **Can required data be input without human intervention?** A Bot does not think; it looks for exactly what it is told to look for and handles what it finds accordingly. If inputs require human judgment, beyond a simple decision of if it can be handled by the Bot, that process is not a good candidate for automation. Tasks that require little to no judgment and have low exception rates are good candidates for RPA.

    For example, an electronic mailbox may receive 300 emails per day, and 100 of them can follow a standard process. It may take a person to manually review the 300 emails and forward the 100 that can follow a standard process to the Bot. But once there, the Bot can take over.

If, however, the mailbox receives 300 emails per day, and each requires that a person read them and look up a variety of different information that may be available from a variety of different and ever-changing sources, then this process would not be a good candidate for automation.

6. **Are potential changes to roles and processes acceptable to management?** One advantage of automating processes is that staff will be freed up for other responsibilities, changing their roles. Also, when the process is being investigated for automation, efficiencies may be determined that will change the process before it is automated. These and related changes must be accepted by the process owner, in order to move forward.

Appendix 5: Risk Assessment Checklist

<Name of project>

Risks	Type	Is There a Mitigation Plan?	Mitigation Plan and Responsible Party

Approvals:

<Name> <Role> _____ Date: _____

<Name> <Role> _____ Date: _____

<Name> <Role> _____ Date: _____

<Name> <Role> _____ Date: _____

Details of the form:

1. **Risks:** In this area, briefly and succinctly identify the risks. One that is standard for RPA initiatives is the following: "The Bot fails to function."

2. **Type:** Generally, the "risk type" is one of the following categories:

   - Compliance
   - Error
   - Financial
   - Operational
   - Reputational
   - Resource
   - System
   - Technology
   - Security

   Complete this area to the best of your current knowledge. Remember, this is a very early stage in the project, and additional risks will be identified as more information about the business need, the technology, etc., is obtained.

3. **Is there a mitigation plan?:** This could be "Yes," "No," or "N/A." "Yes" means that there is a mitigation plan for this particular risk. For the standard risk mentioned earlier, this will be "Yes."

   "No" means that there is no mitigation plan for this particular risk, but one must be created.

   "N/A" means that there is no mitigation plan, but the decision is made to simply accept the risk. This may include such things as accepting maintenance costs, creation of the technical debt, or accepting the fact that X% exceptions will require manual handling.

4. **Mitigation plan and responsible party:** In this box, you will succinctly express what must be done about the risk. If there is a mitigation plan ("Yes" from the column titled "Is There a Mitigation Plan"), briefly describe it. For example, for the standard risk mentioned previously, the "Mitigation Plan and Responsible Party" is usually this: "Revert to manual handling until the Bot is repaired. Business owner and RPA developer."

   If there is no plan ("No" from the column titled "Is There a Mitigation Plan"), briefly describe the steps required to create one. This might include any of the following (among others): "Confer with Corporate Risk Management," "Obtain Input from SMEs," "Research Industry Journals."

> Also in this section, put the role (not the name) of the person responsible for handling the risk. For the standard risk, this is the RPA team lead, along with the role of the person responsible for the process.

At the bottom of the page, include your name and the names of anyone who worked on the risk assessment, the date of the completion of the document, and then the names and roles of the approvers and the date of approval (approvals are generally provided via email, to maintain a "paper" trail).

Remember that the purpose of the Opportunity Assessment is to evaluate the request or RPA candidate for suitability and value prior to investing resources for solution development. Human nature is to jump to execution mode, and it may seem that it is a faster and more productive way, but it is more effective and more efficient to have proper assessment or evaluation of RPA candidates, especially at the early stage of RPA introduction to the organization. The amount of time spent up front will decrease as RPA culture and the technology architecture environment mature. But the introduction of RPA to an organization is not the time to cut corners.

Once all documents required for the Opportunity Assessment phase are complete, you are ready to proceed to the next Governance Committee meeting.

Appendix 6: Opportunity Brief

<Name of process>

Prepared by: <Name>, <Role>, <Department or Business Area>

Current Situation:	Objective:
<At a high level, describe the process as it exists today. Include how long it takes and how many people work on it. Why is the request being made to automate the process? What advantages will automation bring?>	<What, exactly, do you want to automate? The entire process or only certain steps?>
Assumptions	**Constraints**
<What do you assume regarding possible automation. This could be that the Bot will be able to access certain systems, that inputs can be digitized, etc.>	<What do you know will be a challenge to automation? This could be that inputs are not currently digitized or that significant personnel changes requiring human resources input will be required.>
<Include the high-level process map and information about volumes, systems accessed, tools used, and roles currently required.>	

Benefits Estimation:	Proposed Solution
<List the anticipated benefits. This could be reduced headcount, quicker response to customers, more efficient fulfilling of industry regulations, or any other advantage of automating the process.>	<At a high level, describe what the Bot will do.>

Appendix 7: Governance Committee Decision Form

<Name of project>

Decision	Reason for the Decision	Next Steps
<Go, "No Go," or "Wait">	<Why has the decision been made? If "No Go," detail the reasons so the requestor can be advised. If "Wait," describe what additional information is required to make a decision.>	<Describe. For "Go," this will proceed to the next step. For "No Go", it will be to advise the requestor. For "Wait," it will be to obtain the required information.>

Approvals:

<Name> <Role> _____ Date: August 4, 202x

<Name> <Role> _____ Date: August 4, 202x

<Name> <Role> _____ Date: August 4, 202x

<Name> <Role> _____ Date: August 4, 202x

Details of the form:

- **Decision:** This is either "Go," "No Go," or "Wait." "Go" indicates that the governance body has determined that moving to the next phases is beneficial. "No/Go" indicates that for any of a variety of reasons (e.g., insufficient ROI, complexity, etc.), the project will not continue. "Wait" indicates that there is some additional information required before a "Go" or "No Go" decision is made. That will be described in the next column.

- **Reason:** This explains why the decision was made and, in the case of a "Wait" decision, what is lacking.

- **Next steps:** For a "Go" decision, this would indicate the next phase. For a "No Go" decision, this would indicate that the requestor is to be notified and who specifically will notify them.

Appendix 8: Solution Design Document

\<Name of project\>

		Comments
Author		
Date		
Version		
Approval	**Signature**	

Purpose of This Document

- This document contains the solution design for the \<name of process\>. It presents the high-level "As Is" process steps and the "To Be" process steps that will be automated.

- The current process is as follows:

  - \<Include a very high-level description of the process.\>

Goal of the Automation

\<Summarize the goal of automating the process.\>

Applications Involved

Name	Internal/External	Type	Credentials Required?	Read/Write

Inputs

Name	Description	Initial Value

Flow Diagram

\<Include the flow diagram that was created earlier, along with information about volumes, roles, and SMEs.\>

Security and Compliance

\<Include items that are required for security purposes and to adhere to compliance regulations.\>

Process Steps

First Sub-process

Step Number	Description
1	
2	
3	
4	
5	

Second Sub-process

Step Number	Description
6	
7	
8	
9	
10	
11	
12	

X Sub-process

Step Number	Description
13	
14	
15	
16	
17	
18	
19	

<Describe as many sub-processes as needed to convey a thorough understanding of the process.>

Exceptions

- Business
- Technical

Success Criteria

- <Describe what a successful implementation will accomplish, e.g., saving 15 hours per week, increased customer satisfaction, etc.>

Debugging Tips

- <Describe methods of resolving issues that would save time for someone in the future. What are you aware of that could go wrong?>

Disaster Recovery

- <Describe the steps to be taken should the Bot fail.>

Scheduling

- <Describe when the Bot will run, e.g., hourly, daily at 10:00 a.m., weekly, etc.>

Maintenance

- <Describe the regular, periodic maintenance required.>

Appendix

- <Include any documentation required.>

Glossary

- <Include terms that may be unique to the process.>

Appendix 9: Functional Requirements

<Name of project>

1. **Level 1 Details**

   1a. System

   - <What system is the Bot in?>

   1b. Goal

2. **Triggering Event**

   2a. Triggers

   - Pre-condition
   - Post-condition

3. **Level 2 Details**

   3a. Input Parameters

Parameters Name	Description/Value	Mandatory?	Meaning If Omitted (Only Enter If Mandatory = No)
.			

   3b. Output Parameters

Parameters Name	Description/Value	Mandatory?	Meaning If Omitted (Only Enter If Mandatory = No)
.			

4. **Level 3 Details**

   4a. Main Flow Steps

   4b. Alternative Flows

5. **Exception Flows**

6. **Decisions**

Revision History

Author	Version	Date	Comments

Appendix 10: Nonfunctional Requirements

<Name of project>

1. Privacy, Data Retention, and Purge Requirements

Data	Privacy and Retention Requirements	Accessibility	Retention Period
Database			
Technical logs			
Business logs			
Business process data			
Other	<Specify if any other data should be stored.>		
File Share(s)			
Input files	<What files trigger the Bot? What happens to the file then?>		
Output files	<What does the Bot produce? Where are those artifacts stored?>		
Report files	<Does the Bot produce reports? Where are they stored?>		
Other	<Document any other files associated with the Bot.>		
Emails			
Received	<If the Bot reads emails, what happens to them after the Bot has read them?>		
Sent	<If the Bot sends emails, what record of them is kept? Where is it kept?>		

2. **Security Requirements**

Description
What authentication is required for each system and application used?
What roles are attached to the authenticated user?

3. **Disaster Recovery Objectives**

   3a. Disaster Recovery (DR) Designation

   <Identify any system or application required should there be a disaster.>

   3b. Recovery Time Objective (RTO)

   <What is the allowable gap of time (e.g., 2 hours, 24 hours) between a system going down and when they resume at some other site?>

4. **Recovery Point Objective (RPO)**

   <What is the acceptable time for lost data? Date received prior to the last backup will be transferred to the alternate site.>

5. **Scalability (Growth and Volume)**

   <Document any requirements related to scalability. Also, document the expected volume that the Bot can handle, and if there is growth anticipated, note what the anticipated future volume will be.>

Revision History

Author	Version	Date	Comments
.			

Appendix 11: Deployment and Release Setup Checklist

<Name of project>

Complete this template by filling in the information described in the "Item" and "Comments" boxes.

Item	Item Details	Comments	Status*
Product (Bot) Name		Provide the name used to refer to the solution/automation. This name will be used in inventory, reports, change requests, etc.	
Deployment date and time if applicable, agreed upon		As per project plan, provide the planned/preferred date for the deployment.	
Size of automation		It could be small, medium, or large.	
Bot schedule		Specify the preferred schedule (times and days of the week). The acceptable range of start times for the Bot. The approximate runtime of Bot. Are there any service-level agreements associated with the process that affect the scheduling of the automation?	
Contacts during warranty		During warranty, the deployment team may need to contact a software developer on the delivery team that is familiar with the code. Ensure that the developer will be available for five business days after the deployment date.	
Statement of segregation of duties		Delivery team will note names of software developer(s) who wrote the code. Automation services will note names of who user acceptance tested the code and who on automation services is deploying the code. To ensure segregation duties, automation services will ensure that • The software developer who wrote the code was not the same person who performed user acceptance testing. • The deployer of the code to production is not the same person who wrote the code.	
Volume of work		This would be per hour, day, week, etc., depending on the process you are automating.	
Access verification		Mention what type of access is required and if it has been verified in production environment.	
Email addresses for communication		Email addresses of business and technical contacts that will be contacted for any information, errors, or other reasons.	
Disaster recovery		Backout steps, post-implementation steps.	

Item	Item Details	Comments	Status*
Major/minor version		Will this be a large-scale (as described by your organization) change, or a smaller one?	
People who will be present at deployment		Self-explanatory.	
COE contact		Self-explanatory.	
Folder structures		List any new folders required.	
System/application accessibility			
Infrastructure capability			
Risk identification and roles responsible			
Deployment instructions/steps Task deploy steps • Pre-implementation steps • Implementation steps • Test the implementation steps • Backout steps • Post-implementation steps		If applicable, provide detailed playbook for each of the five stages of the deploy. Typically, the Automation Services Deployment Manager specifies the implementation and backup steps.	
Communication channels established		Include a very brief description.	
Production support model		Are you retaining break-fix responsibilities for this automation? If yes, provide remedy support queue name and email contact for the business technical support team. If no, automation services will engage the development team as part of the D2P process to ensure a smooth transition to maintenance knowledge transfer.	

"Status" could be "complete," "incomplete," or "not applicable." For "incomplete" or "not applicable," be sure to include a brief, explanatory comment.

Appendix 12: Process Design Document

\<Name of process\>

1. Description \<Describe the function of the process. Why is it done? How often? And so on. This can be obtained from the Opportunity Brief.\>

2. Current process \<Describe how the process works today; this can be obtained from the Opportunity Brief.\>

3. Systems the process uses \<This information can be obtained from the Opportunity Brief.\>:

   a. \<System 1\>

   b. \<System 2\>

   c. \<System n\>

4. Flow chart of the process \<This was created earlier.\>

5. Detailed process steps \<This was also created earlier.\>

6. Exceptions \<This information was identified during process mapping.\>

Created by:

\<Name, Role, Date\>_____

\<Name, Role, Date\>_____

\<Name, Role, Date\>_____

Approved by:

\<Name, Role, Date\>_____

\<Name, Role, Date\>_____

\<Name, Role, Date\>_____

I

Index

A

Agile methodology, 26, 73, 111
   advantage, 141, 142
   disadvantages, 143
   requirements, 141

B

Bots
   auditing, 115, 116
   categorization, 113
   change management, 114, 115
   checklist, 122
   COE, 98, 99
   communication, 114
   deployment
      access, 101
      automation size, 100, 101
      business, 104, 105
      COE, 103
      development teams, 109, 110
      documents, 106
      emails, 101
      identification, 102
      instructions/documentation, 103
      items, 107
      production, 106
      release options, 103, 104
      requirements, 106
      risks, 103
      scheduled date, 105
      scheduling, 101
      sharing, 106
      software version, 102
      stakeholders, 106
      technical team, 107
      update/enhancement, 102
      validation, 103
      volume of work, 101
      warranty, 105
   disaster recovery, 117
   environments, 106
   LOB/COE, 109–111
   maintenance, 111, 112
   organizations, 98, 99
   prioritization, 112, 113
   release setup checklist/deployment,
      107–109
   retirement
      auditing, 118
      automation
         software, 117
      performance, 118
      process, 117
      steps, 118, 119
   RPA program, 98, 99
   software automation
      tools, 106
   tasks, 120–122
   technical team, 99
   termination, 119
   warranty, 113

BPM Resource Center, 34

Business Process Management Software
   (BPMS), 35

© Robert Fantina, Andriy Storozhuk, Kamal Goyal 2022
R. Fantina et al., *Introducing Robotic Process Automation to Your Organization*,
https://doi.org/10.1007/978-1-4842-7416-3

C

Capability Maturity Model, 1

Center of Excellence (COE), 25, 39, 72, 98–100, 114, 125

Client address and phone number update
access team A backlog, 42
assign request, 42
confirmations, 44
CSS, 39
discovery phase, 42
Email, 43, 44
ID, 43
information, 42
mainframe system/applications, 44
process flow, 40
process map, 40, 42
request form, 40, 41, 43
team's leadership, 40

Client service specialist (CSS), 41, 60

Continuous delivery (CD), 150

Continuous integration (CI), 150

Customer service specialist (CSS), 39

D, E

Decentralized decision-making model, 126

Development methodologies
Agile, 139
business case, 137
challenges, 153
email notifying, 140
roles, 138
RPA, 154
SMEs, 139
user acceptance testing, 140

DevOps
benefits, 151
CD, 150, 152
definition, 149
framework, 150
infinity loop, 151
RPA, 152

F

Framework, operating model
aspects, 24
delivery team, 21

developer, 22
flexible, 27
management and support, 23
path, 23
process analyst/designer, 21, 22
solution architects, 22
well-structured, 26

G, H

Governance, 15, 27
opportunity assessment
impact analysis, 18
process flow, 18
risk analysis, 19
suitability analysis, 19
opportunity identification
process discovery, 18
request form, 17
solution deployment/maintenance phase, 19
solution design phase, 19
solution feasibility decision process, 19
solution life cycle, 19, 20
solution retirement, 19, 20

I, J, K

Initial preparation, RPA
mid-level management idea
early adopters, 11
gains/drawbacks, 11
informal conversations, 12
information, 12, 13
senior management decision
automating tasks, 9
automation, 9
benefits, 10
industry standards, 8
members, 8
opportunity, 9
organization, 9
sources, 7
technology leader, 13, 14

Intelligent character recognition (ICR), 159

Intelligent document processing (IDP), 158, 159

L, M

Land mines, 167

Lean software development
Agile, 147–149
definition, 144
principle, 146
tools, 145
TPS, 144

N

Nonfunctional Requirements
(NFRs), 70, 110

O

Opportunity assessment
address/phone number change
process, 59, 61–64
components, 48, 49
current-state process map, 50
decision, 47
definition, 45
feasibility, 51–53
Governance Committee's decision-
making, 57, 58
risk assessment, 54–57
RPA, 45, 46

Opportunity identification
process discovery phase
approach, 35
BPM Resource Center, 34
BPMS tools, 35
business processes, 36
circumstances, 38
constraints and risks, 39
decision-makers, 34
definition, 34
draft, 37
organizations, 35
process, 38
process map, 36
SMEs, 36
video it, 37
request form, 31
components, 32, 33
requestor, 34

Optical character recognition
(OCR), 159

Organizational models
centralized, 24
decentralized, 25
Hub and spoke, 24

Organizational structure
Capgemini, 133, 134
centralized model, 125, 126
decentralized model, 127–129
hub/spoke, 130–133
operational culture, 135

P

Process design document (PDD), 66, 77

Process mining tools, 157, 162

Product development methodologies
Agile, 26
Waterfall, 25
Waterfall-Agile hybrid, 26

Project sponsorship, 16, 27

Proof of concept (POC), 126

Q

Quality assurance (QA), 85

Quality control (QC), 85

R

Return on investment (ROI), 174

Robotic Process Automation (RPA), 123
automating processes, 2
automation, 166
benefits, 2
bot, 2, 167, 174
definition, 173
development methodologies, 175
early adopters, 165
easy management acceptance, 165
functions, 124
governance board meetings
architect, 168
automation, 169
business, 170
discussion, 168
documentation, 167, 170
document package, 167
email, 168
information, 169

Robotic Process Automation (RPA) (*cont.*)
    overview, 167
    process, 169
    requestor, 168
    wrong tool, 170
  implementation, 12
  initial preparation (see Initial preparation,
    RPA)
  practical applications, 3, 4
  scalability, 176
  tools, 1
  tool selection, 123, 124
  wrong processes to automate, 166
RPA as a Service (RPAAS)
  advantages, 156
  definition, 155
  hyperautomation, 159–162
  intelligent processing, 158, 159
  process mining, 156, 157

S

Solution design document (SDD), 76
  automation, 76, 77, 87, 89
  bot scheduling, 89
  code review, 82
  development/implementation, 78–81
  disaster recovery, 89
  exception messages, 90, 91
  governing body, 66, 73, 74
  high-level determination, 65
  maintenance, 92
  planning, 72, 73
  process flow, 70, 71
  proof of concept/prototyping, 75
  requirements, 67–70
  RPA, 65
  screen captures, 71
  SDD, 86
  technical feasibility, 74, 75
  testing, 82–85, 93, 94
Sponsors, 20, 21
Subject matter experts (SMEs), 35, 68

T, U, V

Technical team, 99, 112
Toyota Production System (TPS), 144

W, X, Y, Z

Waterfall methodology, 25, 65, 137
  benefits, 143
  disadvantages, 144

Printed in the United States
by Baker & Taylor Publisher Services